Leckie ✕ Leckie
Scotland's leading educational publishers

Success guides

Standard Grade
Geography

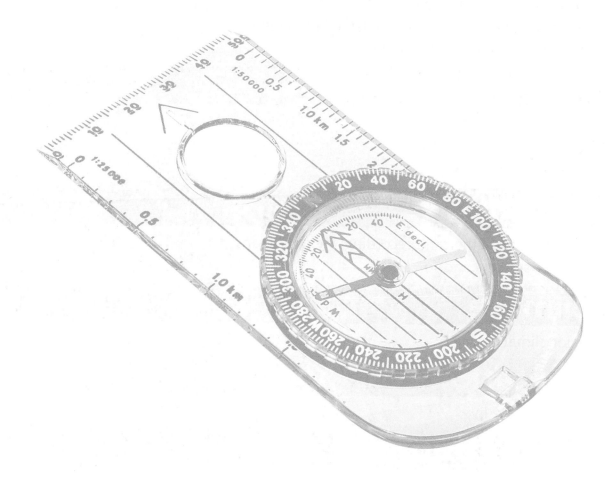

✕ Richard Goring ✕

Contents

Your geography course

In your Standard Grade geography course, you will cover a wide variety of topics. These are broken down into three main areas, all of which will be tested in your examination.

Aspects of PHYSICAL geography

This section examines such things as landforms, the processes which shape the landscapes we see today, weather and climate, how people use the countryside and environmental issues.

Aspects of HUMAN geography

This section concentrates on settlements, rural changes both at home and abroad, farming, industry and economic change.

International Issues

This is an area of geography which looks at the differences populations face across the world. This includes a study of Economically Less Developed Countries and the problems which they have to overcome, and the ways in which trade, aid and international co-operation affect people.

Your final award in Geography is made up of two elements: **Knowledge and Understanding** and **Enquiry Skills.**

Knowledge and Understanding

As the name suggests, this tests your understanding of geographical terms and examples. The ability to explain what you have learned or what you have been given in the exam question is an important part of this element.

This element accounts for 40% of the examination.

Enquiry Skills

This element tests your ability to reach conclusions, to consider different points of view and to make predictions. It also tests your ability to select the best ways of gathering and processing information when carrying out an investigation and being able to justify your choices.

This element accounts for 60% of the examination.

How to use this success guide

This success guide is designed to help you revise for the examination and prepare you to answer questions to the best of your ability. Much of the book is concerned with covering the **key ideas** of the Standard Grade Geography course. It does not tell you everything you need to know for the exam but it will tell you what you need to study in order to get the best result. Read the book alongside your class notes and any relevant textbooks you have, and use it to direct your study. Most pages contain **Top Tips**. These are designed to make you think about important points which will help you in the final examination.

All sections of the book contain **Quick Tests**, which are designed to check that you have understood the topic. Once you are certain of your competence in each topic, you can attempt **Exam-style Questions** which will give you practice for the real examination. Model answers for each of these questions is provided on pages 92–94, so you can check your answers and, if necessary, work out how to improve your answers.

This part of the book is designed to help you study more effectively. It is divided into three parts:

● Preparation for study   ● How to study   ● Some examples

First of all, studying is hard work, although some people find it harder than others. While this book should help you study effectively, it cannot do your work for you! You will find that many of the comments and ideas found in this section will also be appropriate to your other subjects.

Preparation for study

1   Work out a timetable for each week's work in advance.

2   Leave time within this timetable for any homework you may be given.

3   Don't study every night (or day during the holidays). Take some time off (e.g. Friday and Saturday evenings). This gives you something to look forward to as well as an opportunity to relax and socialise.

4   Don't feel guilty on your 'official time off'!

5   Arrange your study time in regular spells, with a break between sessions, e.g. 6.00–7.00, 7.15–8.15, 8.30–9.30. Try to start your studying as early as possible each evening, while your brain is still alert. The longer you postpone working, the more difficult it is to start!

6   Study a different subject for each one-hour spell, except for the day before an exam.

7   During your 15-minute breaks, do something different: have a drink or listen to some music. But don't let your 15 minutes creep into 20 or 25 minutes!

8   Some people study best in silence. No one can study effectively with the radio or TV on. If you must listen to music when you study, keep it at a low volume and of a type which is not intrusive. Keep your passion for heavy metal for breaks between sessions.

9   Have your class notes and any textbooks available for your revision handy as well as plenty of blank paper, a pencil, etc. It's a good idea to have a jotter for study notes so you can refer back to them later. Make 'Catchword' sheets, like the example below – they are invaluable. The blank 'Catchword' sheet on page 95 will get you started.

Catchword	Meaning
Anticyclone	An area of high pressure
Secondary industry	Industries which manufacture things
Commuters	People who travel into town to work
Erosion	The process of wearing down the landscape

10 Finally, forget all or some of the points in this section if you are happy with your present way of studying. Don't change what is successful!

How to study

There are two main types of revision:
- Revision of class notes, worksheets, etc.
- Tackling past-paper or exam-style questions, which can be marked by your teacher.

This success guide can help you with both of these.

Top Tip
Use this book to break your studying into topics, and to direct your study. Go through each of the pages carefully referring back to your class worksheets, answering any questions which are asked, and writing down the answers as you go. You can then refer back to these notes/catchword sheets at a later date for final revision before the exam.

Revising

When you are revising your class work, it is pointless to simply copy your notes out again (how boring!). Set out the topic under revision and note down all 'Catchwords' with a brief description of each. This can best be done on the 'Catchword' sheet. Any points which you think are important should be noted on blank paper, along with any rough copies of maps and diagrams which might be helpful. Don't spend ages colouring in maps – this is merely time-wasting!

Practising questions

Past-paper and exam-style questions can be approached in two ways.

- Under exam conditions, i.e. you can set yourself a time for each question, and answer it as well as possible without using any references or notes. This will encourage you to answer within the time limit. The length for each question depends on the level of paper you are sitting:

  for Credit: about 15–20 minutes for each full question;

  for General: about 10–15 minutes for each full question;

  for Foundation: about 5–10 minutes for each full question.

- Alternatively, you can answer past-paper questions as if they were an ink exercise, using notes, textbooks, etc., to produce a model answer. Your teacher will mark these for you and point out any mistakes or omissions.

Transferring your knowledge

Learning your geography for the exam should not be a case of trying to memorise everything you have done in third and fourth year. Much of the course concentrates on case studies or examples (farms, industrial areas, glaciated uplands, cities in Economically Less Developed Countries, for example). In the exam, you will see questions about places you have not studied. Don't panic! You should be able to transfer your knowledge of a topic or theme to a new area. For example, many of the landscape features which are studied in the Scottish Highlands can be transferred to the Scandinavian mountains, or the Alps, or even the Rocky Mountains in Canada. Economic, social and environmental problems characteristic of declining traditional industries in Lanarkshire are equally true of north-east England or north-east France. And so on.

Some examples

Sometimes, of course, it is necessary to learn or memorise facts. There are different ways of doing this.

Using mind maps

Mind maps are an incredibly useful study tool. You can use them both as a means of presenting facts and other information logically and in a way that is easily understood, and also as a means of revising for tests and exams.

Building up a mind map showing, for example, different types of land use in a city allows you to develop ideas and previously learned information.

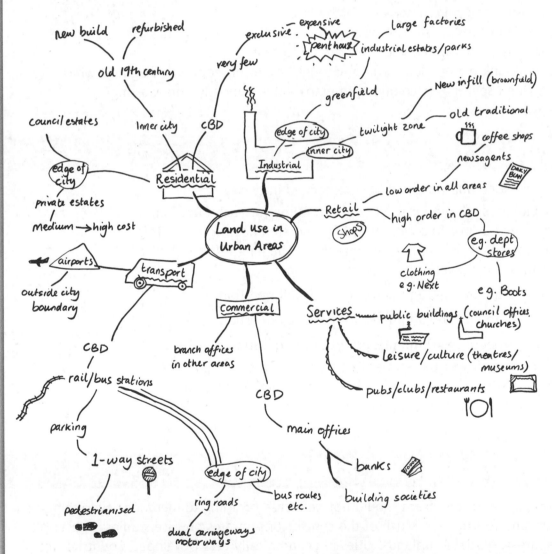

Simple mind maps can also be invaluable for revising, either at the early stages, when constructing a mind map allows your brain to process the information in a structured way, or as a last minute reminder of facts you are sure of. When tackling an extended answer in the exam, a couple of minutes spent sketching out your answer as a mind map helps you remember all the things you need to say, and keeps you on track as you write down your answer.

Using mnemonics

Mnemonics help you to turn information to be learned into something you can easily remember. For example, try arranging the first letters of a list into a sentence.

Scotland's five New Towns could become:

Every **K**ing **C**an **L**ive **I**n **G**lasgow
(**E**ast **K**ilbride, **C**umbernauld, **L**ivingston, **I**rvine and **G**lenrothes)

or Scotland's four major power stations:

Coal **H**eats **T**he **L**ongest
(**C**ockenzie (coal), **H**unterston (nuclear), **T**orness (nuclear) and **L**ongannet (coal))

or The points of the compass in order:

Never **E**at **S**oggy **W**eetabix
(North, East, South and West)

Top Tip
When you are 'designing' a mnemonic, use some words which are familiar, such as football teams, local towns or people's names.

Mapping skills

What kind of map can you expect?

Papers at all three levels will contain a series of questions based on a map extract. Different **scales** and types of map can appear, although the most likely are Ordnance Survey 1:50 000 and 1: 25 000 maps. Tourist maps (e.g. 1:250 000) are possible at Foundation Level.

You will not have to memorise map symbols, as these will always be provided. There is no guarantee, however, that the map extract will be British.

Top Tip
The map extract you get in the exam will include a detailed key with symbols. Take time to refer to this; different scales of map have different symbols.

What skills will you need?

You will be tested on your ability to use a number of mapping skills to extract information from the map, make deductions and identify a wide range of geographical features.

All this may sound rather daunting, but if you can do everything described on the next few pages, you have nothing to fear!

Top Tip
It is essential to practise your mapping skills, so ask your teacher to set you questions on a sample map extract.

Grid references

You will need to use grid references to locate places on any map. These will be 4 figure references at General and Foundation Levels, but you may have to use 6 figure references at Credit Level.

You may also be asked to use the scale of the map to calculate distances, and be able to describe location in terms of direction (for example: '4 kilometres north-east of Stirling').

Contours

Contour patterns can be used to describe **slopes** and **landforms**.

When contour lines are close together, it indicates a steep slope; when they are well spaced, the land is gently sloping. No contours indicate flat land.

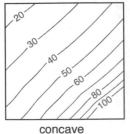
concave

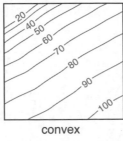
convex

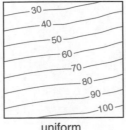
uniform

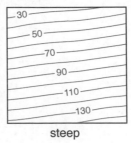
steep

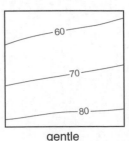
gentle

Quick Test

Many landforms can be identified by examining the contour patterns. How many of the following list can you recognise?

Upland area	Cliff	Headland
Mountainous area	U-shaped valley	Corrie
Plateau	V-shaped valley	River rapids
Ridge	Hanging valley	Waterfall
Pass	Arête	Floodplain
Scarp and dip slope	Pyramidal peak (or horn)	Drumlins
Lowland area	Col	Gorge

Cross-sections

Cross-sections are often used in exam questions. The letters on the cross-section refer to features or locations on the Ordnance Survey map. They will look something like this:

Top Tip
Normally a list of possible answers is provided from which to select responses. Practice makes this an easy exercise.

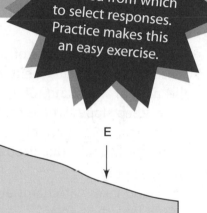

Grid reference

Grid reference

Land use

Different types of land use can be identified from a map. Here are some hints as to what to look for:

- **Pastoral farming**
  Isolated settlements (farms) in an upland or poorly-drained area.

- **Arable farming**
  Isolated settlements in lowland or well-drained areas.

- **Industrial areas**
  New industrial areas are often found on the edge of towns, with a regular road pattern, whereas old industrial areas are normally found near town centres or rivers/canals. They often have irregular street patterns and may have railway lines nearby.

- **Areas of derelict land or wasteland**
  Evidence in the form of spoil-heaps or pit bings are clues to the former industrial use of an area. Urban wasteland is often found in the inner city where traditional industries or poor quality housing has been demolished. Proximity to railway lines is an additional clue.

- **Residential areas**

  Old housing areas are found around the town centre, with closely packed streets, often at right angles. New housing areas are normally closer to the edge of the town, the most modern having gently curving street patterns, often following contour lines, with crescents and cul-de-sacs. In many cities and towns, residential areas have been built recently in reclaimed, former industrial or low-cost housing areas in the inner city. Street patterns are often a clue to identifying these.

- **The Central Business District**

  The CBD or town centre can usually be identified by the number of churches, public buildings (e.g. town hall, museums), bus and railway stations and converging roads.

- **Recreational areas**

  These are often marked on an OS map (e.g. Country Park, golf course) or can be inferred from close observation (e.g. hill-walking in upland areas, bird-watching in woodland areas).

- **Woodland**

  There are different types of woodland that can be recognised on a map. These could be shelter belts which are thin strips of trees near to farmland and natural woodland, often mixed types of trees beside rivers or on steep slopes where they are retained to reduce soil erosion. Commercial forests are often large areas of conifers with access tracks and straight boundaries.

- **Water supply**

  Lochs and lakes with dams are often reservoirs for water supply for urban areas.

- **Other land uses**

  On open, exposed areas, it is often possible to identify uses such as wind farms or radio/television masts.

The identification of different types of land use can be quite difficult, and requires practice.

Top Tip

Why not buy a local OS map and look at areas you know well?

Top Tip

Include at least one grid reference in your exam answer, but don't waste time putting in too many, as there is usually only 1 mark available for this.

Exam-style Question

1. The area in the map extract is typical of a glaciated upland area in Scotland.

   I. Match each of the features below to the correct grid reference.   3KU

   Arête; pyramidal peak; hanging valley; misfit stream; corrie

   977385; 980405; 990420; 003427; 990424

   II. Explain how **one** of the features listed in (I) above was formed. You may wish to illustrate your answer with diagrams.   4KU

2. It has been suggested that Glenshant Hill (9939) would make an excellent site for a wind-turbine farm. Using map evidence, describe the advantages and disadvantages of building such a scheme.   6ES

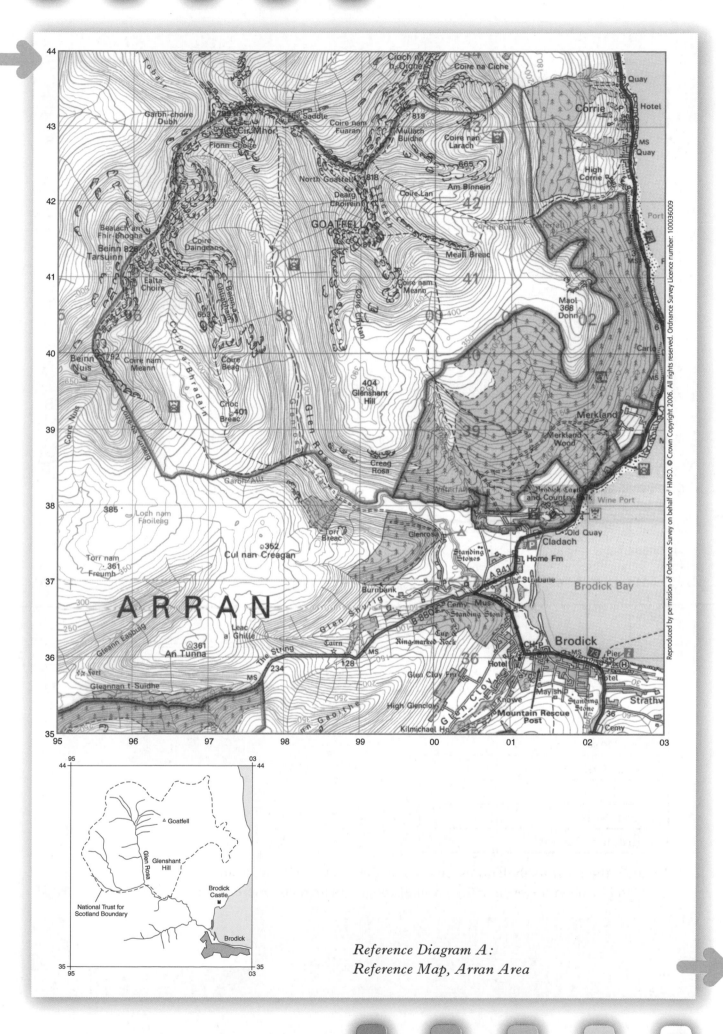

Reference Diagram A:
Reference Map, Arran Area

3. *'The area around Goatfell in Arran would be ideal for development as a large-scale tourist resort. The benefits to the area would outweigh any disadvantages.'*

   Look at the text, Reference Diagram A and the map extract.
   Do you agree with the view expressed in the text?
   Using map evidence, give reasons for your answer.   6ES

4.

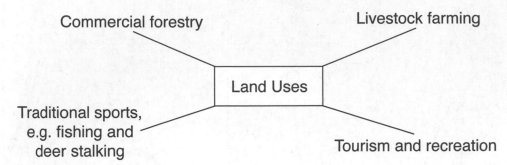

Reference Diagram B: Main Land Uses in the Glen Rosa/Brodick Area.

   Describe some of the conflicts which may occur in a rural area such as the
   Glen Rosa/Brodick area.   6ES

5. Describe the course of the Glen Rosa Water and its valley from its source (966426)
   to its mouth (012366)   4KU

6.

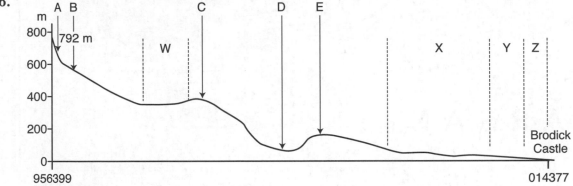

   Complete the table below using letters A–E   4KU

Map feature	Letter
Glen Rosa	
Creag Rosa	
Cnoc Breac	
Beinn Nuis	
Coire nam Meann	

7. Identify the most likely land use at each of the locations W, X, Y and Z.
   Parkland or open forest; marshy ground; rough grazing; commercial forestry.   2 KU

8. Using the sketch and the map extract, name the labelled features.   3KU

Letter on sketch	Name of feature on map
A	
B	
C	
D	

Top Tip
Questions like this are often poorly done in the exam, so the following 3 steps should help you in your answer.
1. Use a clearly identified grid reference in the question to locate the viewpoint on the map;
2. Orientate the map (turn it around) so you are looking at it in the same direction as the view;
3. Identify features, considering whether they are to the right or left of the view, or closer or farther away than the other features.

Exam-style Question

1. Name the grid square which contains Hamilton's CBD (Central Business District). Explain why you chose this square.   3KU

2. Earnock (6954) and Orbiston (7359) are two very different residential areas. Describe the differences between these two residential environments.   4KU

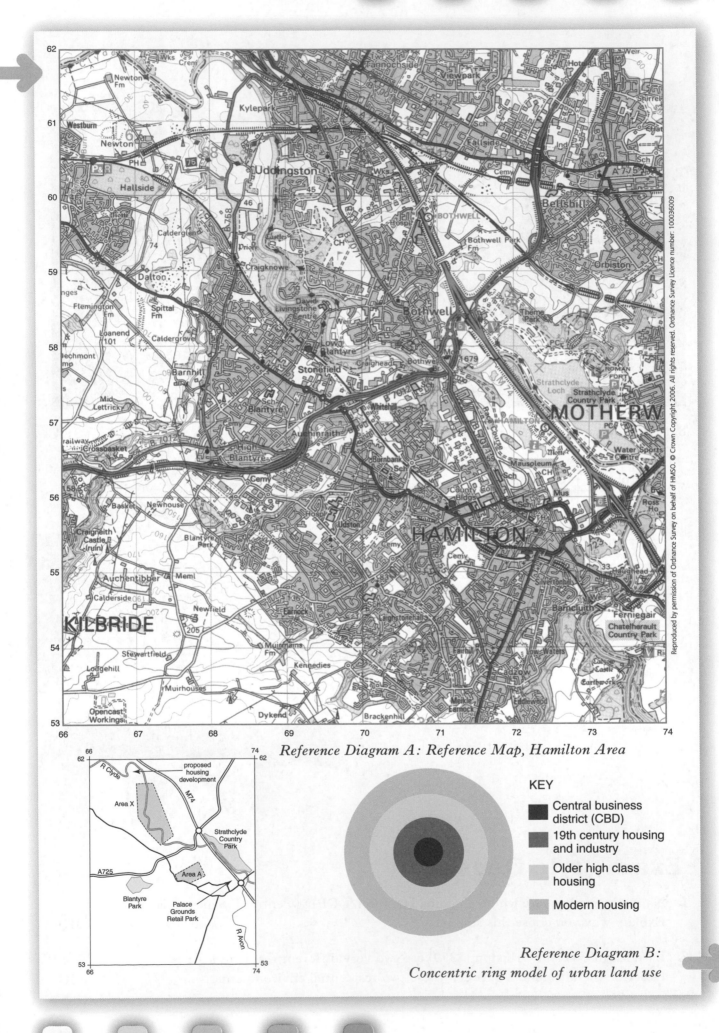

Reference Diagram A: Reference Map, Hamilton Area

KEY

- Central business district (CBD)
- 19th century housing and industry
- Older high class housing
- Modern housing

Reference Diagram B:
Concentric ring model of urban land use

3. A group of students is about to gather information about urban and industrial change in the field study area marked on Reference Diagram A.

   Describe in detail the gathering techniques they might use to complete such an assignment. Justify your choices.

   5ES

4. A technology park has recently been developed at Blantyre Park (6855), as shown in Reference Diagram A.

   Using map evidence, explain the advantages of this site for the both the business community and local community.

   6KU

5. Look at Reference Diagram B which shows a model of urban land use. Using both the model and the map extract, describe the similarities and differences in land use between the model and the land use of Hamilton.

   6ES

6. Describe the physical features of the River Clyde and its valley from 740559 to 668610. 6KU

7. Strathclyde Country Park (see Reference Diagram A) has developed as a recreational facility for many living in the map area. Give the advantages and disadvantages of this site for the creation of a country park.

   6ES

8. *A dormitory settlement is one where the majority of the inhabitants travel to work in a larger settlement.*

   Pupils from a local school wish to find out if Bothwell (7058) is a dormitory settlement. What techniques could they use to gather relevant information?

   Explain your choice of techniques.

   5ES

9. Suggest the type of farming most likely to be found at Muirmains Farm (6854). Give reasons for your choice.

   4ES

10. Using map evidence, explain how physical features (relief and drainage) have affected land use in Area X on Reference Diagram A.

    6ES

11. There is a proposal to build a new housing development to the south of the River Clyde in square 6861 (see Reference Diagram A).

    Would you support this planning application? Using map evidence, give reasons for your answer.

    4ES

12. In what ways has the route of the M74 presented difficulties to engineers between 700610 and 740550?

    4KU

13. A new retail store park has recently opened at the Palace Grounds in Hamilton (729558) (see Reference Diagram A).

    I. Do you think this is a suitable location for a retail park? Explain your answer.

    4ES

    II. Describe TWO techniques which could be used by a research student to find out if this choice of location is a success? Explain your choices.

    4ES

Top Tip

Manage your time carefully when answering the OS map question. It's all too easy to spend too long on these questions and run out of time later in the paper.

Forms of data extraction

Resources in questions

Most of the questions you will answer in the exam are 'resource based'. This means you will be given a map, diagram, sketch, photograph (or a combination of these) from which you will be able to extract information. It is essential to look carefully at all the data you are given. Nothing is there just for show: everything is of use in answering the question.

Sometimes you will be required to change this information from one form into another. This might include:

- Using statistics contained in a table to complete a graph
- Constructing a weather plot, given the relevant information
- Describing the characteristics of a population pyramid
- Measuring the distance between two places on a map
- Describing the course of a river on an Ordnance Survey map.

Sometimes you will have to extract data from a source. This might include:

- Explaining the patterns of land use on a map of a farm
- Describing the trends shown in a graph or table
- Explaining why settlement growth in a map area has been restricted
- Explaining the distribution of steelworks in an area, given the appropriate information about raw materials, transport, markets, etc.

Credit Level questions are likely to have more sources of information. This means you will have to select the relevant information more carefully. The resources you are provided with in a Credit question are also likely to be more complex. However, questions will be worth more marks to compensate.

Top Tip

Take care when completing a graph in the exam. Use a pencil and ruler so any mistakes are easily corrected. Be precise – marks will be deducted if you are careless or inaccurate.

Answering knowledge and understanding questions

Geographical knowledge

Questions testing knowledge are fairly straightforward to answer. They often require you to identify or select examples of geographical concepts.

This type of question might include:

- Name three examples of features of glacial erosion found in the OS map.
- In which grid square would you find the CBD?
- Identify the river features found in the block diagram.
- Describe the likely social and economic effects of the decline of steelmaking in Lanarkshire.
- Describe the distribution of arable land in the map area.

Geographical understanding

Questions testing understanding are rather more difficult. Not only must you be able to identify geographical concepts, you must be able to explain them.

For example:

- Explain, with the use of a diagram, how a corrie is formed.
- Explain why the city centre grew up in grid square 1243.
- In what ways have the river features shown in the sketch restricted the development of settlement beside the river?
- Explain why traditional industries like shipbuilding have declined over the last fifty years.
- Explain the distribution of arable crops in the map area.

When answering questions where you are asked to "explain" something, it is essential to make sure you don't simply describe.

If, for example you are asked to "explain the distribution of arable crops on a farm", you will not receive marks for simply describing the different crops found in different areas. You need to say **why** they are found where they are. This might be "wheat is found on the gentle slope facing south to receive maximum sunshine for ripening and above the flat valley floor where soils can be waterlogged, whereas root crops like turnips are more tolerant of damp soils so they are planted nearer the river."

When asked to **explain**, you will know you are answering properly when you use words like "because", "therefore", "the reason is" in your answer.

Questions testing understanding are usually worth more marks than those testing knowledge.

As you can see, answering questions testing knowledge and understanding depends on the care and attention you have put into studying your class notes. This book is designed to help you channel your study into the right areas. It cannot possibly hope to cover all the course content, but, used systematically with your class notes, it should ensure that you will be able to answer every question with a high hope of success.

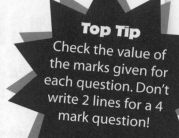

Top Tip
Check the value of the marks given for each question. Don't write 2 lines for a 4 mark question!

Answering enquiry skills questions

Enquiry skills

Questions which test the skills of enquiry are rather different from those testing knowledge and understanding. They do require an understanding of the basic geographical concepts, but it is your ability to make informed judgements which is most important. Answering these questions takes practice, and it is essential that you are aware of exactly what you are being asked to do. Enquiry questions will almost always provide you with sources of information which you will have to examine before coming to conclusions.

Making a decision

Sometimes you will be asked to make a decision based on information received. For example:

- Using map evidence, describe the advantages and disadvantages of the site of Haddington.
- Explain which types of foreign aid shown in the diagram would most benefit Uganda.
- Using the information given, do you think the Cruachan site was a suitable location for a hydro-electric power station?

Making a choice

Sometimes you will be confronted with a number of points of view, and your task will be to support or reject each one. This requires careful handling: make sure all the points of view have been considered. In effect, you are being asked to examine both sides of an argument and decide which is most suitable. For example:

- The area shown on the map is to be made into an industrial estate. What are the advantages and disadvantages of this plan?
- Argue the case for and against the proposed quarry.
- Explain how the development might bring benefits and problems to the area.

In the exam, you will get marks for answering both points of view. When asked to make a choice, you should select the answer you can find most points in support of, and then, if you feel you haven't written enough, you can add "On the other hand....", making points in support of the alternatives.

Gathering information

An important aspect of geography is the gathering of information which allows you to make sense of the environment around you. This could involve you in the collecting of **data** from the physical or human environment (such as measuring the depth and speed of flow of a river, or noting different types of land use in an urban area). These are described as **primary sources** because you gather the information first-hand from the environment.

Another way of gathering information is to use **secondary sources** such as textbooks, photographs or maps. Someone else has already gathered the information and presented it in a way you can interpret. In an exam, you might be asked what type of techniques you would use to gather information for a certain purpose and say why you think this technique is best.

Top Tip
Make sure you examine and refer to **all** the sources in the question.

Gathering techniques

Information can be gathered from a variety of sources. These can be primary, secondary or a combination of both. There are many sources of information, and it is normal to use several different types. Sources include:

TEXT   e.g. books, magazines, newspapers, libraries, posters

MAPS   e.g. Ordnance Survey (OS), historical, geology, at different scales, town plans, atlases, soil maps

DATA   e.g. statistics, census reports, Scottish Office, library databases, CD-ROMS, graphs, tables

ILLUSTRATIONS   e.g. photographs, sketches, cartoons, field sketches, diagrams

FIELDWORK   e.g. visits to streets, settlements, factories, rivers, upland areas, farms and a million other places!

At both General and Credit Level, you will have to say which technique would be best to gather information for a particular topic and **justify** your choice by saying why, in your view, it is the most suitable.

There are many different techniques which can be used to collect information:

Observation

This is the main source for the geographer. Not only is the information gathered up to date, you can make sure it is directly relevant to a topic of investigation. Here are some examples:

- Measuring maximum/minimum temperature, air pressure, wind speed and direction over a period of time to record changes in the weather
- Identifying different types of farming land use to explain the relationship between land use and relief
- Measuring the depth, width and speed of flow of a river at different points along its course
- Carrying out traffic-flow surveys to explain traffic movement
- Measuring river pollution at different points along its course and explaining the differences

- Sampling natural vegetation at different altitudes using a quadrat and explaining the findings in terms of soil, moisture levels and aspect
- Observing the impact of walkers on hillside routes
- Carrying out a survey of order of services in different settlements, or different areas of the same settlement and explaining your findings.

Questionnaires

A questionnaire is a useful way of gathering information from other people. Care needs to be taken to make it easy to complete and relevant to the investigation topic. It can be used for:

- Asking shoppers about their shopping habits and origin to identify order of service and the sphere of influence of a centre
- Investigating travel to work patterns
- Researching the sphere of influence of a leisure centre or recreational facility
- Finding out about holiday trends.

Interviews

While a questionnaire is best for asking a few simple questions of a large number of people, an interview is more detailed and can be used to gather in-depth information from one or two people. Interviews could be used when:

- Asking a farmer about changes to his/her farm over a period of time
- Finding out from a park ranger what problems arise from public access to a nature reserve
- Asking a factory manager about the location factors affecting his/her industry
- Finding out from a senior citizen about changes in a town or landscape or industry over the last fifty years.

Top Tip
Know the differences between a questionnaire and an interview. Be able to say WHO you are interviewing or asking a questionnaire of, and give examples of your questions.

Sampling

Because there is so much information out 'in the field', geographers make use of **sampling techniques**. This means a random or controlled selection of places or individuals in order to collect representative data. For example, it is impossible to question all the shoppers in a town centre, or record all the vegetation on a sand dune, or count all the traffic on a road in twenty-four hours, or measure river speed at all points on its course.

When sampling is chosen as a technique, it is important to make sure it is representative. A traffic survey, for example, of four roads leading into a town centre should be carried out at the same time on each of four days (not weekends); otherwise comparisons cannot be made. Similarly, a questionnaire of shoppers should not be directed only at attractive seventeen-year-olds, as this would give a biased result!

Field sketching

This is a very useful technique which allows you to record relevant information as you see it. It is particularly useful for buildings (shops, housing types, factories) and landscapes (valleys, farms, glaciated scenery). Those of us who are not very artistic might take photographs instead, and use these as a basis for sketching once back home or in the classroom.

Maps

Maps are an invaluable source of information, especially large-scale maps such as 1:10 000 or 1: 25 000. These can show a great deal of information about land use and can also be copied and used as the base map for fieldwork observations.

It is also interesting to compare maps of different ages: this may show changes in industry, housing, transport networks and rural land use. Remember too that there are many different kinds of maps. Geology maps tell us about rock type, and there are maps showing vegetation, land quality, soil type, climate, population densities and tourist information.

Processing information

Once information has been collected, whether from a primary or a secondary source, it needs to be **processed** in order to make it understandable. There are many ways of doing this (graphs, charts, maps and annotated sketches are examples) and you should be able to say which is best for a particular purpose.

Identifying appropriate techniques for processing information

Once information has been gathered, it needs to be **processed** in a way that makes it easy to understand and use.

There are many different techniques for processing information:

Drawing graphs

Different types of graphs can be used for different purposes.

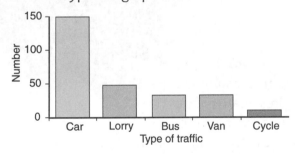

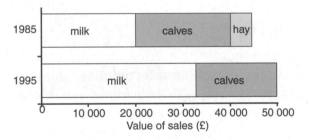

Bar graphs are useful for data which can be broken down into different parts (e.g. traffic data).

Another type of graph is a divided bar graph which is good when showing relative proportions (e.g. farm sales).

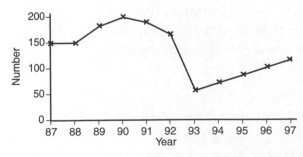

Line graphs are used when data shows a trend over time (e.g. sales of milk calves from a farm).

Top Tip
A common weakness in many candidates' answers to processing questions is to reproduce learned answers which do not relate to the data provided in the source for the question. Make sure you do refer to the source information!

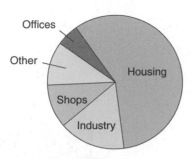

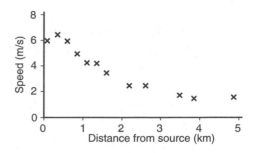

A pie-chart is best used when data can be represented as percentages. One example would be land use in a town. A trend over time can be shown using several pie-charts in sequence.

A scatter graph is a good way of showing how two factors relate to each other (e.g. speed of river flow and distance from source).

Drawing maps

Maps can be used to show all kinds of information gathered in the field. A map can be drawn from scratch (showing, for example, a farm's land use) but in most cases information will be added to a base map which is copied from an OS or other map.

Maps can be used for many purposes, the most common being to show land use (urban or rural), changes in land use (e.g. the growth of a settlement), sphere of influence, distribution of observed data (e.g. corner shops), transport routes, areas or zones (e.g. high ground, areas prone to flooding, zones for industrial development) and location (e.g. New Towns in Scotland, cattle markets or football grounds).

Drawing cross-sections and transects

The use of a cross-section or transect is an excellent way of visually relating observed data to another factor such as altitude or distance from a known point. One example would be a cross-section showing the altitude of a farm and relating this to land use.

A transect is more useful when relating land use to a known point (for example, distance from the CBD).

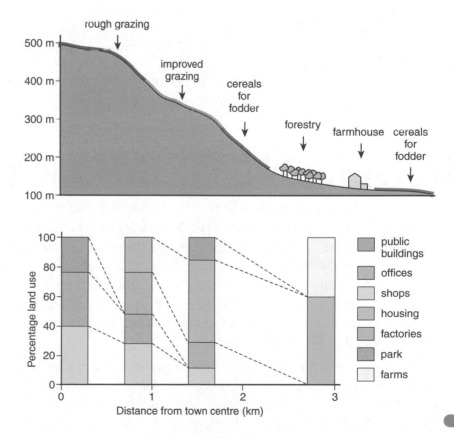

Annotating

Both field sketches and photographs can be annotated (labelled) so important details can be identified or recalled. Houses, for example, can have information added about building materials (local stone, brick, slates, tiles), window types, alterations, state of repair; factories can have details showing age (e.g. number of storeys, tall chimneys, soot-blackened brickwork, lack of parking space, cramped surroundings).

Classifying information

Information gathered during fieldwork can be very complex and varied, and it is often a good idea to **classify** it so that sense can be made of it. A range of types of shops found in a town centre, for example, can be classified into groupings such as clothing stores, food shops, entertainment, financial services, fancy goods and furniture stores. This allows comparisons to be made more easily with other centres or other parts of the town. In the same way, crops observed during a farm study can be classified into cereals, root crops, fodder crops, market garden produce and so on.

Top Tip
You can practise annotating by using old photographs or cuttings and adding information in arrowed boxes.

Information can also be put into a **table** – this allows easier comparisons to be made from the data collected. One example could be a table of weather readings made over a period of time:

Day	Pressure (mb)	Wind direction	Wind speed	Minimum temperature	Maximum temperature	Cloud cover
Monday	1020	NE	5	7	21	none clear sky
Tuesday	1000	SE	2	8	17	$\frac{1}{4}$
Wednesday	990	SW	15	9	14	$\frac{3}{4}$
Thursday	985	SW	22	10	15	overcast

Another way of representing data is in the form of a matrix, where it is easier to show relationships between different variables.

How many marks?

General questions are usually worth a maximum of 4 marks, and Foundation questions are seldom worth more than 3 marks each. In General and Foundation papers, you are given spaces to answer in, so you have a good idea of the length of the answer which is expected. Questions in the Credit paper call for more extended answers, as you would expect. They are normally worth between 3 and 6 marks. In the Credit paper, you will write your answers in an answer booklet so try to match what you write to the marks available. Don't write 20 lines for a 4 mark question!

In the exam, 1 mark is given for a simple point and 2 marks for a developed point. For example: "The factory is situated close to major roads" would be worth 1 mark, whereas "The factory is situated beside major roads for the easy transport of raw materials and finished product" would be worth 2 marks.

Processes

Shaping the landscape

All physical landscapes are shaped by a number of **processes**, which act on the landscape and cause it to change. The most important of these processes are **erosion**, **transportation** and **deposition**.

Erosion

Erosion is the process which eats away at the landscape, gradually wearing it down.

Physical erosion occurs when rock is broken up into smaller pieces by the effect of ice, waves, rivers or wind.

Transportation

Once rock has been eroded, it is **transported** away by water, ice or wind.

Deposition

Eventually the material which has been eroded and transported is dumped or **deposited** somewhere new. This creates new landscapes.

Weathering

Weathering is very important, as this is the process which softens up the landscape, making erosion much easier.

One example of weathering is **frost-shattering**. This happens when the temperature falls below freezing point. Water which is caught in cracks in the rock freezes and expands; this allows pieces of rock to become loose, weakening the rock and making it easier for other forces to remove it and cause erosion.

Gravity also has a part to play in weathering: pieces of rock which fall off cliffs collect at the foot of the cliff in a **scree slope**. **Landslides** are also caused by gravity acting on soil or rocks which are unstable – often the presence of water lubricates the slope and it begins to slide.

Top Tip
Unlike erosion, which implies that eroded material is removed from its original location, weathering occurs in situ.

Top Tip
Frost-shattering is most noticeable in temperate climates, as temperatures cross the 0°C threshold most often. In very cold climates, where temperatures are almost always below freezing, very little freezing and thawing takes place so frost-shattering is not a major cause of weathering.

Time

All these processes act on the landscape over a period of **time**. Sometimes the period of time in which change occurs is very short, as in the case of a landslide. Usually, however, the timescale is much longer – in the case of glacial erosion, it may be tens of thousands of years.

Rock type

The type of rock (or **geology**) in an area is also important in controlling the rate of erosion. Resistant rocks, such as granite, wear down much more slowly than soft rocks, such as chalk. This is why most of the mountains in Britain are composed of resistant rocks; they have resisted erosion more than the softer rocks around them which now form the lowlands.

Exam-style Question

Look at the diagram below.

Explain in detail how a gorge is formed.

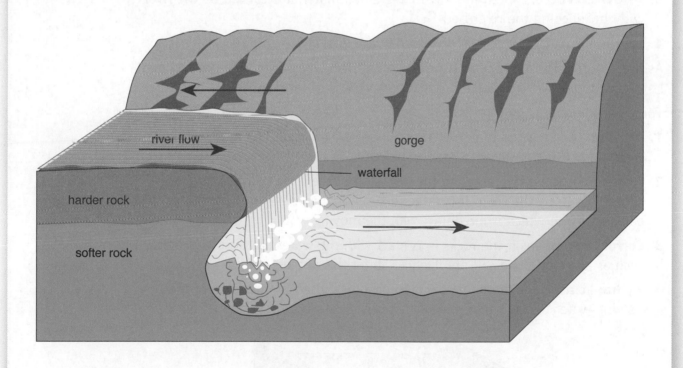

Rivers (1)

River erosion

The main agent of erosion and deposition which is responsible for creating river landforms is **water power**.

When rain falls, or snow and ice melt, water flows downhill under the force of gravity until it reaches the sea or a lake. As it flows, the water picks up fine material and carries it along. When there is a lot of water, or it is flowing very quickly, it can carry a great deal of material. This material is called **sediment** or **bedload**.

This bedload is moved or **transported** by the river in a number of ways:

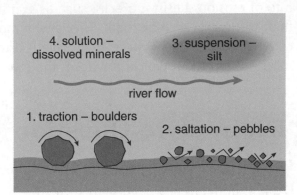

1. **Traction** describes the movement of material by the force of the river – normally by rolling or sliding along the river bed.

2. **Saltation** happens when material is 'bounced' along the river bed. This usually happens to smaller material, such as pebbles, or when the river is flowing forcibly.

3. **Suspension** occurs when fine or small-grained material is carried by the river without touching the river bed.

4. **Solution** is when material is dissolved in the water.

As this material is transported, **erosion** occurs. This wears away the underlying landscape and carves out valleys, creating a variety of **landforms**.

This process happens in four ways.

1. **Attrition** occurs when material carried along by the river hits into other material and breaks up into smaller pieces.

2. **Corrasion** is the name given to the **abrasion** of the river bed by material carried along by the river. It is rather like sandpaper wearing away at the earth's surface which is in contact with the moving water.

3. **Corrosion** happens if the rocks in contact with the water are dissolved. Rocks such as limestone and salt are likely to suffer this process.

4. **Hydraulic action** is the name given to erosion caused by the force of the flowing water hitting other material. This is greatest during times of flood.

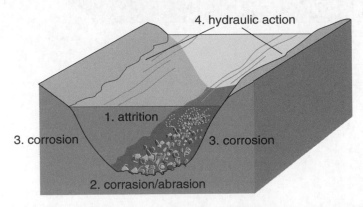

The diagram and table below show the differences in a river in its **upper**, **middle** and **lower courses**. This refers to the different processes involved at each stage, as well as to the typical landforms and the way that the river is used.

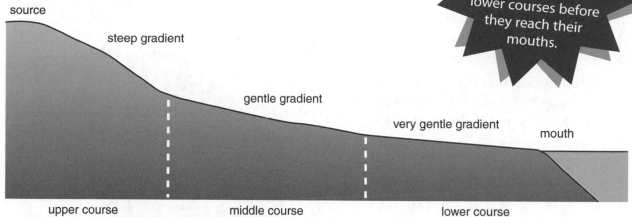

- A cross-section of a river from source to mouth is known as a long profile.
- A river usually has an upper, middle and lower course.
- The gradient (slope) of a river decreases from source to mouth.

Upper course	Middle course	Lower course
mainly erosion	erosion and deposition	mainly deposition
mainly erosional landforms	both erosional and depositional landforms	mainly depositional landforms
upland farming, HEP stations, wind farms, forestry, recreation.	small market towns, mixed farming.	large towns, ports, large-scale industry.

Quick Test

Why does the size of material carried by a river decrease in size from source to mouth?

Answers Sediment carried by the river in the upper stage is broken into smaller pieces as it moves downstream. By the time it reaches the lower stage, most of the material is very fine-grained and easily transported.

Rivers (2)

Upper course stage of a river

The erosive power of rivers is greatest in its **upper course**, where gradients are steepest and there is a large amount of angular bedload to wear away the land.

The landforms typical of the upper course are mainly caused by erosion. These include **interlocking spurs** and **V-shaped valleys**.

As the river makes its way downhill through an upland area, it erodes a path which winds between interlocking spurs. Most of the erosion is at the bottom of the river and as it deepens its course, it creates a valley which is V-shaped in cross-profile. As the river's path becomes eroded, the sides of the valley become unstable and fall into the river, where they are transported downstream, adding to the sediment and causing further erosion. A feature of fast-moving, swirling streams are **potholes** which are deep holes in the bed of the river, created by the grinding (abrasive) effect of rocks trapped in the depression.

Valley sides are weathered, loose material falls into river and is carried downstream.

Hills called interlocking spurs are formed on either side of the river.

steep sides

river erodes vertically

V-shaped valley forms

River begins to flow from side to side.

Middle course stage of a river

In the middle course there is a mixture of erosion and deposition. As the gradient of the river is usually less, there is already a fairly substantial amount of carried sediment and the speed of the river is generally slower.

Changes in rock type and sudden variations in slope often result in rapids and waterfalls, where the fast-flowing water containing high amounts of sediment can create a very effective erosional tool. The diagram shows how a waterfall can recede upstream and leave a scar or **gorge** in high land. When the river reaches less steep land, it slows down and is unable to carry so much sediment. This material is deposited as **alluvium** and the river begins to move in a snake-like pattern. This produces **meanders** where the faster flowing water on the outside of the bend cuts into the alluvium, creating a **river cliff**, and the slower moving water on the inside of the bend deposits more material (sand and shingle) in a beach known as a **slip-off slope**.

2. Harder rock eroded more slowly, forms overhang.

3. Overhang eventually collapses and the waterfall moves upstream.

4. Steep sided valley called a gorge is formed.

river flow

harder rock

softer rock

waterfall

gorge

1. Less resistant rock eroded more quickly, undercutting harder rock.

6. Deep plunge pool is eroded at base of waterfall.

5. Large pieces of rock – remains of previous overhang.

Lower course stage of a river

In the lower course, the work of rivers is almost entirely depositional, although during times when the river is in spate (flooding), erosion can be considerable.

Meander loop becomes very large and inefficient.

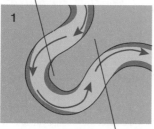

During a flood, narrow neck of land is eroded.

River breaks through and follows the shortest course.

Meander is cut off to form an ox-bow lake

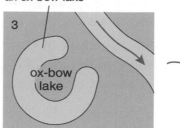

ox-bow lake

Over time the lake becomes a marsh, and then dries up completely.

■ erosion
■ deposition
↘ fastest flow

The river is flowing at its slowest and because it is carrying a lot of fine-grained material, much of it in suspension, it can deposit this sediment easily. The gradient is low, however, and deposited sediment can clog up the river's path, causing it to flood and change its course. Meanders become much larger and sinuous, spreading across a very wide **floodplain**. The erosion of the outer bend of the meander can cause the river to straighten its path and change direction, leaving an abandoned meander. If this becomes cut off from the main river, the water stagnates and an **ox-bow lake** is formed. If the river floods, huge amounts of sediment, mainly sand and muds are deposited and these can create **levées** along the side of the river channel. These levées can help prevent further flooding.

If a slow-flowing, sediment rich river flows into a lake or sea where there is little or no tide, the speed of the water drops drastically and sediment is deposited in the shape of a **delta**.

Top Tip
River **estuaries** such as the Firth of Clyde have lots of flat land near their mouths which is ideal for large industries or port development.

In its upper stage, you will find:

- A narrow V-shaped valley with steep sides
- Boulders and sharp rocks in the river bed.

In its middle stage, you will find:

- A wider valley with less steep sides
- Smooth stones and pebbles in the river bed.

In its lower stage, you will find:

- A wide, flat valley
- Sand and silt/mud on the river bed.

Quick Test

Give **one** example of tourist activity found in each of the stages of a typical river valley.

Answers Examples could include: Upper course – walking, mountaineering, skiing, Middle course – fishing, canoeing, walking, Lower course – sailing, rowing

Glaciation (1)

Glacial erosion

The main agent of erosion, transportation and deposition responsible for the formation of glacial landscapes is **ice power**, although **water power** also produced landforms as the ice was melting at the end of the **ice age**.

Glacier ice is a very effective agent of erosion, as it can gather rocks and debris in its sole (or base) which, as the ice flows across the landscape, grind away at the earth's surface like a sheet of sandpaper. This process is known as **abrasion**.

Rocks can also become embedded in the ice itself, and as the ice moves away, lumps of rock can be **plucked** from the landscape. These processes explain the power of glaciation in shaping the landscape.

Landforms of glacial erosion

Corrie

This is a hollowed-out depression at high altitude in upland areas where glaciation occurred. During the **ice ages**, snow falling did not melt in the summer months, but accumulated over many years on higher ground. As more snow fell, it was squeezed and compressed into ice. The ice collected on the top of mountains and in hollows or corries. This ice began to move (or flow) slowly downhill, eroding the ground with which it had contact. Corries

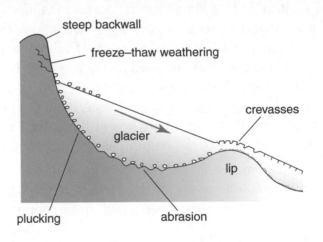

thus grew larger as the ice age continued. After the ice age, a deepened corrie was left, with a steep back wall and lip at the entrance to the corrie. Today, it is common to see a **tarn** or **corrie lochan** at the base of the corrie.

Arête

As the glacier ice was enlarging the corrie by the processes of plucking and abrasion, the back wall of the corrie became steeper as it ate back into the mountain behind it. If a mountain had a number of corries eroding into opposite sides, arêtes formed; corries arranged around a peak eroded back to create a **pyramidal peak** or **horn**.

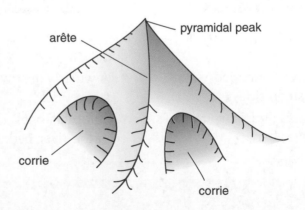

U-shaped valley

As the ice flowed downhill from the higher land, it tended to follow the valleys previously carved out by rivers. The ice was more direct, and eroded the sides of the valley as well as the base, resulting in the formation of a valley that was U-shaped in profile. The moving ice had cut off (or **truncated**) the interlocking spurs. Smaller glaciers joining the main glacier would fall down the valley sides, leaving high (or **hanging**) valleys perched above the main valley. Today, they often have streams of water cascading into the main valley as a waterfall. Many U-shaped valleys contain lakes; they are long and narrow, known as **ribbon lakes**.

In Scotland, as in much of north-west Europe, the rise in sea level after the ice age (caused by the melting ice) meant that many U-shaped valleys were **inundated** by the sea. This resulted in the formation of sea lochs or **fjords**.

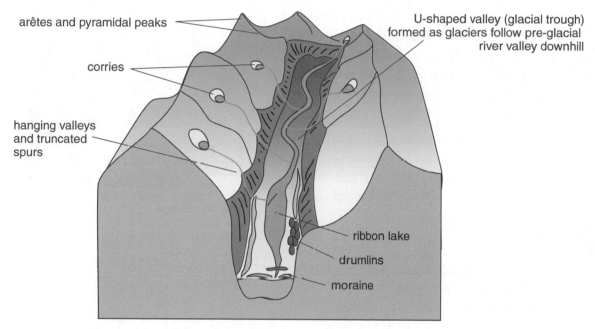

Roche moutonnée

This is the name given to a large piece of rock on the valley floor which has been shaped by the passing ice. The side of the rock facing the ice flow is smoothed whereas the side downstream is very roughly plucked to form a jagged rock surface.

Glaciation (2)

Landforms of glacial deposition

Glacial debris

The erosive power of the moving glacier creates a huge volume of debris which is transported on top of, inside and underneath the ice. Some of this material is moved as the ice moves; a large amount is transported by melted ice or **meltwater**.

This debris is known as **till** (or **boulder clay**) if it is moved by the ice.

Moraine

This is the general name given to debris carried and deposited by a glacier. In mountainous areas, there is much evidence of moraine, as shown in the diagram. **Lateral moraine** is deposited at the edge of the glacier as it moves down a valley; when two valley glaciers meet, a **medial moraine** can be formed. At the end of the glacier (its **snout**), a large amount of material is deposited; this is known as **terminal** or **end moraine**.

Higher up the valley, huge amounts of debris have been left by the glacier; this can be recognised today as **ground** or **hummocky moraine** – it is made up of material ranging from fine sands and mud to huge boulders.

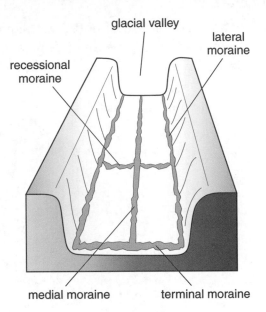

glacial valley

lateral moraine

recessional moraine

medial moraine

terminal moraine

Erratics

Such is the power of the moving ice that large pieces of rock which fall onto the glacier can be transported many hundreds of kilometres and dumped well away from their place of origin. These are called erratics.

Drumlins

These are unusual little mounds of moraine which are found in areas where the glaciers were melting. Although there is some uncertainty about how they were formed, they were clearly shaped by the ice and often occur in large numbers known as **swarms**.

direction of glacier flow

drumlin

Eskers

When the ice was melting, there would have been a huge volume of water flowing over, inside and under the ice. Like any river today, these streams would have carried much bedload. Rivers flowing in pipes through the ice would deposit sediment, leaving a long, often twisting, line of debris behind when the ice melted. These lines of debris are called eskers.

Downstream from the melting glacier, rivers carrying huge amounts of sediment would flow downstream, depositing sands and gravels as **outwash deposits**.

Top Tip
One way of telling the difference between debris carried by glaciers and debris carried by rivers is the shape of the rocks. Glacially moved sediment is usually angular; sediment that is carried in a river has been worn into smooth, rounded material.

Using glacial landscapes

It is very important to be aware of how people make use of glacial landscapes. For example, hanging valleys and U-shaped valleys are suitable for the development of hydro-electric power; glaciated uplands attract hill walkers and mountaineers because of their impressive scenery; corries are often ideal for skiing; many areas of fertile farmland are found on glacial deposits of till or boulder clay. Other activities to consider include forestry, holiday development, army practice and sand and gravel quarries.

Top Tip
When the ice sheets were melting, huge volumes of meltwater carved channels in the landscape and transported massive volumes of debris.

Settlement – tourist resorts such as Aviemore have developed.

Skiing – ski runs have been developed – extensive network of chair lifts and cable cars.

Hydro-electric power (HEP) – steep slopes, high precipitation and snow melt are ideal for HEP.

Forestry – coniferous forests grow below an altitude of 500 m. Wood is used for fuel, building and paper making.

Walking – dramatic scenery attracts tourists on summer walking holidays.

Recreation – watersports and fishing opportunities.

Farming – sheltered valley floor with deep fertile soils – suitable for dairy farming and crops such as hay.

Communications – deep, straight valleys provide natural routeways for road and rail links.

Quick Test

Give six reasons why glaciated upland areas such as the Lake District attract so many tourists.

Answers could include: impressive scenery; combination of hills, forests and lakes; mountains/hills for hill walking, climbing; lakes for water sports; many attractive towns and villages; good access by road; large population within two hours drive

Exam-style Question

Look at the diagram.

Select **one** feature caused by glacial erosion and **one** feature caused by glacial deposition.

Explain, using diagram/s, how each feature was formed.

Elements of weather & recording; weather maps

Weather elements

The main **elements of weather** that you need to be able to recognise are: precipitation, temperature, sunshine, wind speed and direction, air pressure, visibility and humidity.

You should also be able to recognise and describe what weather instrument is used to measure each of these elements, as well as the unit of measurement.

Element of weather	Instrument	Unit of measurement
Precipitation	rain gauge	millimetres (mm)
Temperature	Thermometer	degrees Celsius (Centigrade)
Sunshine	sunshine recorder	hours per day
Wind speed	Anemometer	knots/kilometres per hour
Wind direction	wind vane	north, north-east, east, etc.
Air pressure	Barometer	millibars (mb)
Visibility	human eye	kilometres (km)
Humidity	Hygrometer	% humidity
Cloud cover	human eye	oktas (one okta is one-eighth of the sky covered by cloud)

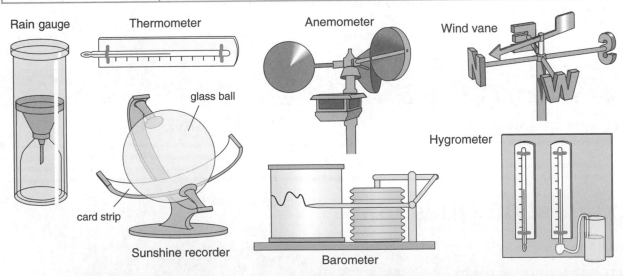

Rain gauge   Thermometer   Anemometer   Wind vane

glass ball

card strip

Sunshine recorder   Barometer

Hygrometer

The weather station

Instruments to measure the weather are found in a **weather station** which is located in an open area to allow the most accurate readings to be made. The site of a weather station must be well away from buildings, trees and walls and should be protected by a fence to prevent animals or curious humans from interfering with the readings.

Weather maps

A weather (or synoptic) chart is a map showing the actual weather recorded at a specific time. Each weather station's weather reading is shown on the chart as a **weather plot** or weather symbol.

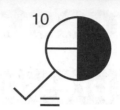

south-west wind, 10 knots, mist,
5 oktas of cloud cover, temperature 10°C

Weather terms

Weather maps can be described using the following terms:

Isobars	lines joining points of equal pressure
Anticyclone	an area of high pressure
Depression	an area of low pressure
Ridge	a narrow area of high pressure between two areas of low pressure
Trough	a narrow area of low pressure between two areas of high pressure
Cold front	boundary between warm air and advancing cold air
Warm front	boundary between cold air and advancing warm air
Occluded front	this is rather like a cold front, formed when a cold front catches up with a warm front.

Top Tip
The weather conditions associated with an occluded front are very similar to those of a cold front.

Quick Test

Describe the weather conditions at the weather stations with the following plots:

a) 4

b) 10

c) 14

Answers a) Northerly wind, 15 knots, snow showers, 2 oktas of cloud cover, temperature 4°C. **b)** Easterly wind, 5 knots, no precipitation, clear sky (no cloud), temperature 10°C. **c)** Westerly wind, 20 knots, rain, sky completely covered (8 oktas), temperature 14°C.

Weather systems and forcasting

High pressure and low pressure

An area of high pressure (or **anticyclone**) is characterised by stable weather conditions. In summer, this usually means warm, sunny weather, with little or no cloud and light winds. In winter, clear, calm conditions are often associated with cold, frosty nights and cold, sunny days, although fog and mist can occur.

An area of low pressure (or **depression**) is linked to cloud, wind and rainfall. In Britain, this usually means relatively mild conditions in winter and miserable weather in summer, with low cloud and hill fog.

The diagram below shows the main characteristics of anticyclones and depressions.

Anticyclone (in Northern Hemisphere) – high pressure

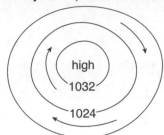

Isobars well spaced – light winds

Winds rotate clockwise

Summer – clear skies, sunshine

Winter – frost and fog

Depression (in Northern Hemisphere) – low pressure

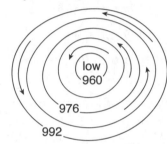

Depressions move quickly Winds are strongest where the isobars are closest together

Winds rotate anti-clockwise

Summer and winter – cloud, rain and wind

Where air masses meet

When two air masses come together, there is an area of unstable air at the boundary which is called a **front**.

A synoptic chart is a representation of the air pressure and fronts as they pass over an area. The depression in the diagram below is typical of many of those passing over Britain. It shows a warm air sector between two fronts, and winds circling an area of low pressure in an anticlockwise direction.

Transposing this information to a cross-section through the warm sector, as shown below, helps to explain the weather associated with the depression on either side of, and between, the two fronts.

Top Tip
Winds blow anticlockwise around a depression in the Northern hemisphere, clockwise around an anticyclone.

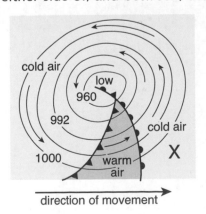

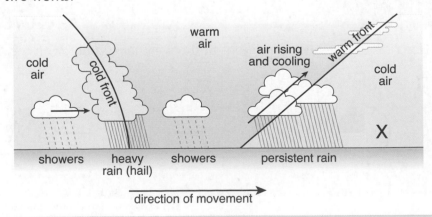

In the examination, you might be asked to describe and explain the weather changes at a location as a depression passes. These diagrams help you to do that. The diagram shows a location 'X' which is about to be passed by the depression. As the depression approaches and passes, the following sequence of changes in the weather would be typical:

As the depression approaches:

- Weather is bright and high level cloud appears. Winds will be light and south-westerly (because winds rotate anticlockwise around a depression);
- Winds strengthen (isobars getting closer) and cloud base falls as warm front approaches. Light rain falls.

As the warm front passes overhead:

- Rain becomes persistent (heavy drizzle is common). Temperature rises and winds increase, becoming westerly (closer isobars and as centre of the depression moves eastwards, the wind direction changes).

In the warm sector:

- The rain eases, although showers occur. Temperatures stay warm and winds still strong.

As the cold front passes overhead:

- Heavy rainfall, sometimes thundery with hail. Winds may strengthen and turn to a north-westerly direction (again, see isobars in synoptic chart). There is a sudden drop in temperature as winds from the north-west (colder, Arctic air) are drawn in.

After the cold front:

- Winds begin to ease and turn to northerly (cold). Sunny with (sometimes heavy) showers.

Exam-style Questions

1 The weather system shown in the diagram to the right is moving in an easterly direction.

Explain in detail what will happen to the weather in Newcastle in the next twenty-four hours. You should make reference to temperature, wind speed and direction, precipitation and air pressure.

2 Look at the diagram below.

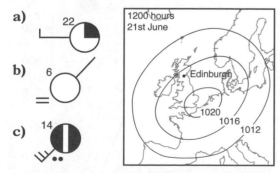

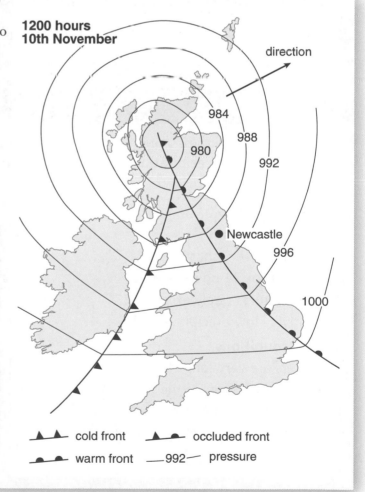

Which of the three weather station plots shows the weather conditions most likely experienced in Edinburgh at noon on 21 June?

Explain your choice in detail.

Major climatic zones

Climate zones to remember

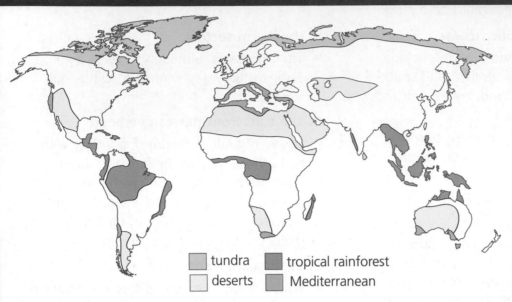

tundra | tropical rainforest
deserts | Mediterranean

Given a world map showing major climatic zones, you should be able to identify and name at least two areas of:

- Rainforest (e.g. Amazon and Indonesia)
- Hot desert (e.g. Sahara and Atacama)
- Tundra (e.g. Alaska and northern Russia)
- Mediterranean climate (e.g. southern Europe and California)

Top Tip
In the examination, you may be asked to draw or complete a climate graph, given the temperature and precipitation figures for each month of the year. Remember that temperature is shown as a line-graph, precipitation as a bar graph.

Graphs

You should be able to describe each of these climates in terms of **temperature** and **precipitation**. A typical climate graph for each of these areas is shown below.

Tundra areas

Summer conditions

- summer is very short
- daylight for 24 hours in north
- temperatures rarely exceed 18°C
- relatively low precipitation

Winter conditions

- winters are very long
- 24 hours darkness in the north
- temperatures often below 20°C
- little precipitation which falls as snow
- strong wind chill

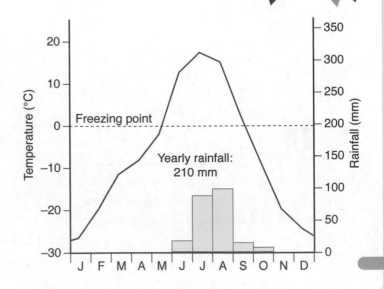

Freezing point

Yearly rainfall: 210 mm

Mediterranean areas

Summer conditions

- generally hot and dry
- very little precipitation
- daily temperatures often reach 30°C

Winter conditions

- generally mild and damp
- temperatures rarely fall below zero
- relatively high precipitation

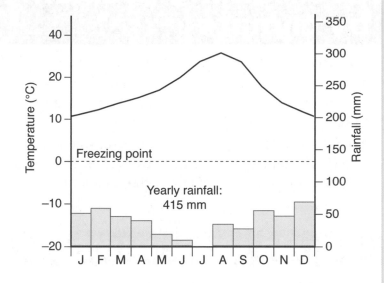

Yearly rainfall: 415 mm

Hot desert areas

Summer conditions

- very hot
- virtually without rainfall
- daily temperatures exceed 30°C

Winter conditions

- similar to summer conditions
- high temperatures
- very low rainfall
- daily temperatures often in excess of 30°C
- in hot deserts, there may be many years without any significant rainfall; when rain does fall, it may be torrential, resulting in flash floods.

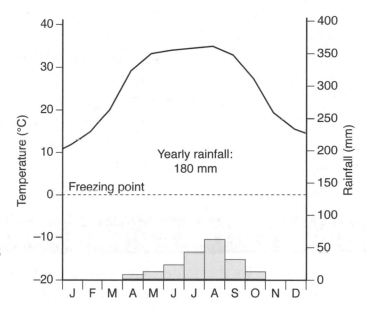

Yearly rainfall: 180 mm

Rainforest areas

Summer and winter conditions are very similar

- very high temperatures, often around 30°C
- high daily and annual rainfall
- very heavy rainfall most days in early afternoon, as strong sun heats the land surface, causing rapid rising of warm air which cools to produce rainstorms
- high humidity

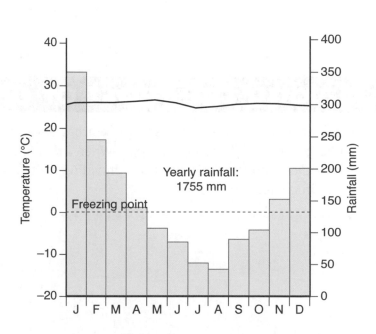

Yearly rainfall: 1755 mm

Overcoming problems caused by climate

For each of the above areas, you need to be aware of:

- The way in which the climate has limited development. For example, high temperatures and rainfall all year round have made areas of rainforest inhospitable and difficult to work in.
- One example of a recent development where, for economic reasons, the difficult conditions have been overcome. This could be oil exploration in the tundra or desert, or mining in the rainforest.

Fragile environments

You should also be aware of some examples of how such recent developments in environmentally sensitive areas can have a serious impact on nature and the landscape.

Examples could include mineral extraction in rainforest areas such as the Amazon rainforest where the impact includes:

- Destruction of the vegetation cover
- Loss of many animals and other species because of loss of habitat
- The need to move local people from their traditional areas
- Considerable change to traditional ways of life and customs.

How climates can make life difficult

You should also be able to describe how each of the climates has created problems for people living in the areas. Places with Mediterranean type climates, for example, suffer from summer drought and high temperatures. Successful agriculture is often only possible using **irrigation** or **water transfer** from other areas.

Making the most of difficult climates

On the other hand, the climates in these areas have advantages too. Many people go to Spain for their holidays. When irrigation is possible, crops can be grown quickly and early in the year, providing fruit, vegetables and flowers to market in cooler climates.

Exam-style Questions

Describe the typical climate of each of the two graphs below.

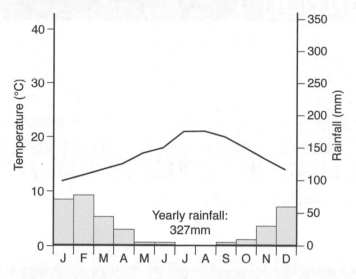

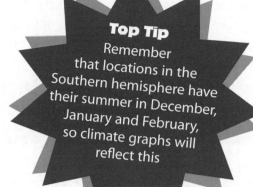

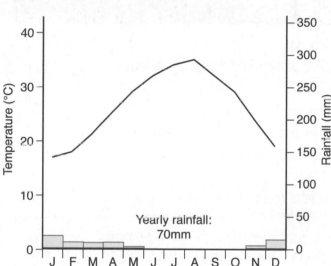

Quick Test

Drought can be a major problem – how could this affect tourism or farming?

Answers Tourism is very important in many areas with Mediterranean climates, as the summer weather is perfect for holiday-making. If there is a summer water-shortage, however, this coincides with the demand created by visitors for hotels (showers, sewerage and swimming pools) and restaurants, bars, etc.
Farmers need to irrigate their crops, especially fruit and vegetables in the summer months. This is needed to provide food for the increased population (tourism) and export.

Where people live

Population distribution 1

Over the centuries, people have colonised most of the world's land surface. Some areas such as the tundra, mountains, rainforests and hot deserts are sparsely populated because of the hostile climate and environment which make survival difficult.

Population distribution 2

Most of the world's **population** is found in **temperate** and **tropical** areas where the climate makes farming easier, and the weather is normally less extreme.

Where the living is easy

As a rule, people choose to live in an area which has:

- A climate that will permit successful agriculture to support the population
- Sufficient water to drink and allow crops to grow
- Soils which are fertile enough to produce crops and allow grazing
- Land which is flat enough to make transport of people, food and goods easy
- A supply of raw materials (e.g. wood, metals, fuel) which can produce wealth.

Obviously not all places have all these requirements to the same extent, but it is generally true to say that the better the supply they do have, the more developed that place is. Temperate areas such as northwest Europe have a good supply of these requirements and so have developed into wealthy areas over the last several hundred years. In many cases, less developed areas such as much of southeast Asia have less of these resources and it has been difficult to produce wealth and so become more developed.

Many agricultural areas in ELDCs have very high population densities because large families work the land in an attempt to produce enough food for survival. Often the land is not very productive, or is over cultivated, and productivity is generally low. In EMDCs, agricultural land supports only very low population densities, because farm produce is sold at market. These highly mechanised farms require few workers and the majority of the population lives in urban areas.

Population growth

With rapid population growth in the last hundred years concentrated in the tropical parts of the world, the pressure on resources has increased. This has meant that there are more people to share what resources there are and so, rather than become more developed as might have been expected, poverty has increased and many countries are now less developed than they were in the past. This can be seen in African countries such as Kenya and Zimbabwe.

Top Tip
Some countries such as Saudi Arabia appear to be very developed, with huge annual wealth, but this can be shared among relatively few people leaving many people very poor.

Quick Test

Describe the natural factors that have led to Britain being an Economically More developed country

Answers A moderate climate, neither too hot, too cold, nor too dry; Land than can support good farming; Relatively flat land allowing easy movement of goods and people; A good supply of raw materials such as timber and metals, fuel and water.

Different ways of using land

Competing for rural land

In rural areas throughout the world, there are many different ways in which the countryside is used. Farming, mining, forestry and tourism are examples of different **land uses**. Often different land uses are in competition for the same area of land, and this can lead to conflict. Farmers, for example, can be in conflict with tourists who want access to their land for walking or sightseeing.

Typical land uses in a rural area

The diagram on the right shows some of the land uses which may be in competition with each other.

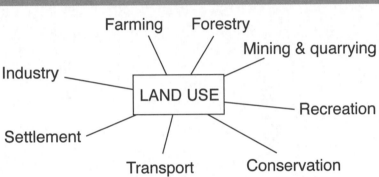

You should be able to describe one example of each of the land uses shown above:

- Recreation could include skiing, climbing, walking, camping, fishing or car-rallying.
- Farming in upland areas is mainly hill-sheep farming or deer or cattle rearing, but many types of farming are found in lowland areas, such as arable farming, market gardening and dairy or meat production.
- Forestry is mainly found in hilly or remote areas; it can be privately owned or managed by the Forestry Commission.
- Mining or quarrying can range from mining for fuels (e.g. coal) to quarrying for rock for cement or road surfacing. This normally leaves unattractive scars on the landscape, and is often opposed by local residents.
- Conservation suggests that an area of land has been designated as an area of particular environmental or scientific interest. Nature reserves, national parks and wildlife sanctuaries are examples. These can include mountainous or moorland areas, coastal sites and river banks.

- Transport is increasingly a major influence on rural areas. New motorways, by-passes and high speed rail links are examples. When a rural area attracts visitors, as in a national park such as the Lake District, the greater numbers need improved transport systems to accommodate them.
- Settlement can range from traditional villages and towns which have existed for hundreds of years to modern developments catering for the increased number of people who want to relocate in the countryside (e.g. pensioners) or buy second homes for holidays. Improved transport networks and more people working from home mean there is great pressure on rural settlements to expand, often at the cost of the local environment.
- Industry ranges from small-scale enterprises such as fish farms to large industries like power plants, paper mills or wind farms. Their impact on the environment can be considerable.
- As the urban population grows, more houses, factories and recreational areas are built at the edge of cities and towns. This is often on existing farmland, posing a threat to rural areas.

When competition becomes conflict

It is also important to understand some of the conflicts which may occur. For each of the land use conflicts shown below, you should be able to describe how the demands of each create problems for the other:

- **Farming v. recreation** – When people visit the countryside, they attract many services such as hotels, restaurants, shops, car parks; these use up existing farmland. Careless visitors can leave gates open, dump litter and interrupt farmers' business.

- **Mining v. farming** – Mining activities such as open-cast coalmining causes unsightly scars on the landscape. Heavy lorries can pollute the environment and disturb livestock.

- **Forestry v. farming** – Planting forests can use up farmland for many years until it is ready for harvesting. Although forests tend to be planted on more hilly land, the water running from forests can be acidic which can damage soils.

- **Industry v. recreation** – Many factories are ugly and affect the visual quality of an area, as well as cause noise, pollution and increased traffic, all of which can harm potential recreational activities.

- **Industry v. farming** – industry in rural areas can create problems similar to mining, as described above.

In the exam you may be asked to give the points of view of both sides in the dispute. For example, for industry v. farming:

> *U-shaped valleys in the Scottish Highlands are often ideal for the production of hydro-electric power, because they are easily dammed to give reservoirs and there is high rainfall which provides for a steady supply of water. Farmers in the area, however, are against such developments because the land which ends up under water is the lowest and most fertile land in areas where there is very little land suitable for farming. Often farmhouses are flooded out and although farmers receive compensation, their future is limited to less profitable hill sheep farming on the higher land.*

Top Tip
There are always two sides to a conflict. Rehearse answers for and against different land uses.

Exam-style Questions

The Lake District is a popular destination for many holiday makers. Do you think all the land users in the diagram will welcome the visitors?

Explain your answer.

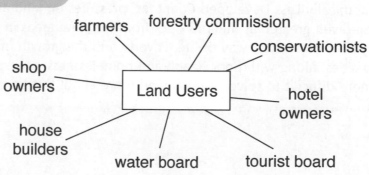

farmers forestry commission conservationists

shop owners

Land Users

hotel owners

house builders

water board tourist board

Conservation and urban expansion

More and more people

As world population grows and cities, particularly in the **Economically Less Developed Countries (ELDCs)**, become even larger, it is inevitable that more and more countryside becomes swallowed up by built-up areas.

Pressure on resources

What makes the problem even worse is that there is a greater urban population requiring land for houses and factories, while at the same time requiring more food from the countryside. And of course, this food is often grown on land which the cities are growing into!

Restricting urban growth

In the **Economically More Developed Countries (EMDCs)**, **green belts** have often been created around cities in an attempt to prevent **urban sprawl**. You should be able to name at least one city (e.g. Dundee or Edinburgh) where there is a green belt, and explain some of its implications. These could include **higher land values**, **leapfrogging** and **breaching**.

Land values rise as more people want to buy houses near the countryside but still near the city. Leapfrogging occurs where certain developers at the edge of the city are refused permission to develop, so nearby towns or villages attract potential house buyers. Breaching is the name given to an area which runs through the green belt but in which development is allowed. This can often happen beside a major road or near existing industries.

Exploding cities!

In Economically Less Developed Countries, pressures on land around urban areas are even greater, as millions flock from the rural areas in search of employment or a better way of life. Poverty and the growth of **shanty towns** around cities, along with poor health and homelessness, make the problems even more difficult to solve. You will find more about this on page 57.

City problems

The main land use pressures around most cities in the Economically More Developed Countries are:

In the examination, you may be asked to describe why each of these exerts pressure on the countryside around an urban area.

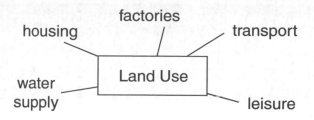

housing · factories · transport · water supply · Land Use · leisure

For example:

- As more people want to live in attractive areas, there is continuing pressure to build on the farmland at the edge of the city. Farmland is easy to build on, and house buyers are willing to pay a lot of money for being close to the countryside. Of course, as more and more houses are built, older houses get farther from the edge of the city!

- Industries are always keen to build on greenfield sites, as there is plenty of room for building and car parks, and access to main roads is easy. For example, developers of retail parks are keen to set up at the edge of large towns and cities.

- There is usually traffic congestion in city centres and planners are eager to by-pass urban areas by building ring-roads and motorways. The farmland at the edge of the city is ideal for this. The increase in air transport also means there is a need for new airports or extra runways – again the green belt around cities is seen as ideal.

- Leisure facilities are increasingly looking to build at the edge of cities. Access is easy, and there is room for building. Golf courses, football grounds and sports centres are good examples.

- Cities need a supply of drinking water and many reservoirs are located around existing urban areas. Expansion of the urban area can threaten the quality of water available.

There are both benefits and disadvantages in urban expansion at the rural/urban fringe. Some local people will be upset by more building, traffic and the loss of rural access. Others may see the advantage of better communications and more services. In an exam, you may have to consider both sides of the argument or decide whether this kind of development is a good thing.

Top Tip
You should have a knowledge of an urban centre (e.g. Glasgow or Edinburgh) and be able to describe some of the pressures caused on the urban–rural fringe by different land uses such as housing developments, out-of-town shopping centres or transport requirements. It is essential to be able to give actual examples when answering questions.

Quick Test

Look at the diagram on the right.

With reference to a city in an Economically Less Developed Country you have studied, describe the pressure each of the land uses puts on the countryside around the city.

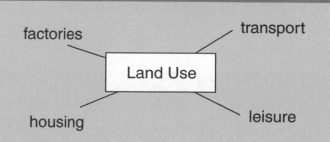

factories · transport · Land Use · housing · leisure

Answers Water supply – land for reservoirs for expanding population; Housing – huge demand for land – both legal and illegal – for incoming people; Industry – cheap labour costs attracts industry; often on city edge; Farming – need to provide food for growing population.

Our green planet

Planet Earth can be described as an **ecosystem**. This means that it is composed of many millions of living things (plants and animals) which depend on each other for their survival. If there is damage to any of these living things, it will have an effect on others, and there may be far-reaching consequences.

Using the Earth's resources

There are also many non-living things on Earth. These include **resources** such as minerals (e.g. iron and oil) and timber. Some of these resources are **renewable**. This means that they will naturally replace themselves if they are given the time to do so. Timber and water power are examples of renewable resources. Other resources are **non-renewable**, and as people use them up, they will become scarcer until they eventually run out. Metals and fuels such as coal and oil are examples of non-renewable resources.

A sensible use of resources

As the population of the world continues to grow, and people expect an ever-improving standard of living, it is inevitable that resources will be used up at a faster rate. The only solutions that will overcome this problem are:

- to reduce the rate at which we use up resources
- to attempt to recycle as many resources as possible
- to find alternatives to scarce resources
- to adopt policies of **sustainable development**.

Top Tip
Study one example of a resource that is being over-used and think about problems that may occur if it continues to be exploited. Examples could be oil, fish or copper.

Quick Test

Explain what is meant by recycling. Name five common products which can be recycled.

Answers Recycling is re-using a product once it has been used. Examples could include paper, glass, plastic, iron (cars), aluminium (cans), mobile phones and copper wire.

Conservation in sensitive areas

In some parts of the world, the impact of resource-gathering is especially damaging.

Damaging our planet

In forest areas, particularly **rainforest**, the cutting down of forests for timber or farmland has major local implications as well as threatening the global environment.

Removing the forest canopy exposes the fragile soils to the effects of the weather – rainfall washes the soil away and makes cultivation difficult. Productivity falls and farmers find it increasingly difficult to make a living. On a larger scale, there is the loss of valuable plants and traditional products such as timber, nuts and rubber. The reduction in forest means less carbon dioxide is changed into oxygen and it is feared this will have a global impact. You should be able to describe these effects, and explain the global consequences of forest clearance, tying them in to what is happening elsewhere on Earth. You should be able to describe the effects on **global warming** and the implications of the **greenhouse effect**.

Global warming is the name given to the gradual increase in the average temperature of the Earth's atmosphere, oceans and land surface. This has happened on many occasions in the Earth's history, but there are concerns that the present-day global warming is caused largely by the consumption of fossil fuels which produces gases that absorb heat in the atmosphere. Power stations, cars, factories, homes and animals are all responsible for producing greenhouse gases.

The greenhouse effect is a natural feature whereby the atmosphere, because it contains certain gases, mainly carbon dioxide and water vapour, retains a proportion of the heat of the sun. This keeps our planet relatively warm and at an even temperature, so allowing plants to grow and animals, including humans, to thrive. Without it, our planet would have extremes of temperature which would make life unlikely.

Desertification

In **arid** (dry) areas, increasing pressure on land caused by changes in the climate patterns and increasing population leads to land degradation and eventual **desertification**.

Desertification can be caused by a combination of factors including:

- **Climate change** – global warming has had the effect of reducing rainfall in many parts of the world, particularly those areas where farming is marginal and a period of drought can have serious consequences.

- **Over grazing** – Increasing population in semi-desert areas has forced people to keep more grazing animals such as cattle and goats. This has in turn led to less surface vegetation, turning grasslands into deserts.
- **Population growth in rural areas** – more people means greater pressure on the land in areas where there is a great dependence on agriculture. This can lead to soil erosion and a reduction in land capability.
- **Deforestation** – as explained above, the removal of tree cover encourages rainfall to wash soil into rivers with disastrous effects on farming potential.

Tackling desertification is complex, because the factors which cause it are not easily solved.

Countries where soil erosion is a major concern are often Economically Less Developed, so their governments cannot afford either to stop cutting down rainforest or to provide jobs for local people so they have less need to grow their own crops and raise their own cattle – all of which results in damage to the land.

Some of the most successful ways are by teaching local people how to work as a community to save and use water wisely.

Local people are now encouraged to replant trees to prevent soil erosion, and use drought resistant crops where possible. This can help stabilise the soil and prevent further damage by wind or animals. Practices such as contour ploughing and terracing can slow the rate of water run-off, again reducing soil loss. The digging of deep wells and creation of irrigation channels can breathe life into otherwise threatened environments. The Economically More Developed countries such as Britain can help by supporting Fairtrade schemes or investing in Aid programmes which tap into local knowledge and experience. This is explained further on page 88.

Polluting our water resources

Oceans and lakes have also suffered greatly in the last 100 years from the dumping of waste, over-fishing and pollution.

For many centuries, it was believed that the seas and oceans were so large and deep that any waste dumped in them would be so diluted as to render it harmless. Over the years, a cocktail of **effluent** including sewage, industrial waste, chemicals and radioactive waste has been dumped in the oceans and lakes of the world. The water becomes polluted and may kill off wildlife such as mammals, fish and seabirds.

Even more vulnerable are enclosed seas like the Mediterranean, or inland lakes such as the Great Lakes in North America or the Caspian Sea in Russia which cannot flush themselves out and so poisons and toxic waste become more concentrated.

Modern trade involves the movement of vast amounts of goods around the world by sea. Tankers carry oil, natural gas, uranium, iron ore and many manufactured products many thousands of kilometres, and if storms or human error bring disaster, pollution is the inevitable result. The oceans break up toxins very slowly; the effects of spillage may last for many years.

These effects can be reduced by a number of measures, including international rules on the sea – worthiness of vessels, better training for ships' officers, routing vulnerable cargoes away from dangerous environments, improved weather forecasting, and the increased use of pipelines for transporting oil, gas and chemicals.

The problems of over-fishing

The continuing growth of the world's population puts pressure on oceans, seas and lakes in a different way. Fish and other marine life are a valuable food supply, but a global market encourages over-fishing and fish stocks are becoming increasingly **depleted**. The North Sea is a classic example. The European Union has consistently reduced quotas for fishing fleets; now many sea-going fishermen have had to sell their boats because they can no longer make a living.

The effects of damaging our environment

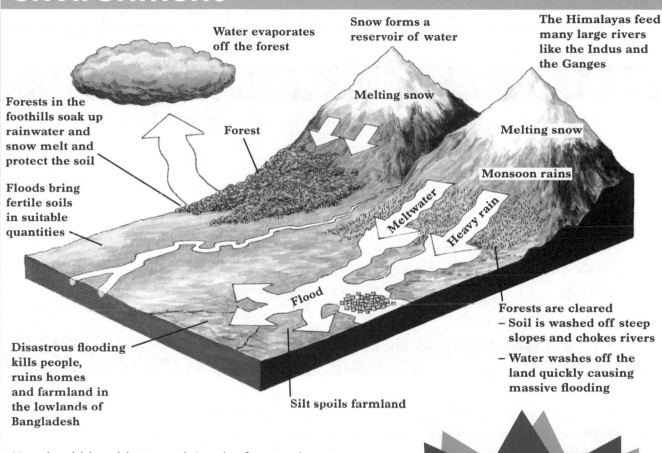

Water evaporates off the forest

Snow forms a reservoir of water

The Himalayas feed many large rivers like the Indus and the Ganges

Forests in the foothills soak up rainwater and snow melt and protect the soil

Floods bring fertile soils in suitable quantities

Forest

Melting snow

Melting snow

Monsoon rains

Meltwater

Heavy rain

Flood

Disastrous flooding kills people, ruins homes and farmland in the lowlands of Bangladesh

Silt spoils farmland

Forests are cleared
– Soil is washed off steep slopes and chokes rivers
– Water washes off the land quickly causing massive flooding

You should be able to explain why forests, deserts and oceans are so liable to be damaged. This means you will have to think about the local ecosystems and local people, as well as the global effects of damage.

The diagram above shows some of the effects of forest clearing in Bangladesh.

Top Tip
Remember that whenever part of the natural environment is damaged or destroyed, there is always an impact on other parts of the environment. Deforestation in upland areas, for example, will increase run-off and may encourage flooding. Wildlife and human activity will also be affected. A mind map is a good way of showing and remembering this.

Quick Test

What is the greenhouse effect?

Answers This is when energy (heat) from the sun is trapped in the earth's atmosphere.

What is a settlement?

Settlement size

A **settlement** is a place where people live and work. The size of a settlement can vary from a single building (e.g. a farm) to a major city such as London with over eight million inhabitants.

Depending on their size, settlements can be described as:

HAMLETS (smallest)

VILLAGES

TOWNS

CITIES

CONURBATIONS (largest)

Can you name examples of each of the above?

What do we find in urban areas?

There are many different **land uses** to be found in settlements. These include:

RESIDENTIAL (housing)

COMMERCIAL (shops and offices)

INDUSTRIAL (factories)

RECREATIONAL (parks, sports facilities, etc.)

WASTE LAND (derelict buildings, cleared land)

TRANSPORT (roads, railways)

You should be able to identify each of these on an Ordnance Survey map.

Urban land use

Very often, the different land uses identified above are found in certain parts of a settlement. The shops and offices are usually to be found near the centre of the settlement. In large towns and cities, this commercial area is often called the **Central Business District** (or **CBD**).

Quick Test

1. There are many more small settlements than there are large ones. Can you explain why?

2. Look at the Ordnance Survey map of the Hamilton/Bellshill area on page 16. Match each of the following grid references to one of the land uses in the green box above.

   A 696588 **B** 723565 **C** 733603 **D** 682562 **E** 725592 **F** 688595 **G** 736538
   H 675588 **I** 685620 **J** 722563 **K** 704603 **L** 703583 **M** 732577

Answers 1. Small settlements can occur anywhere; large settlements have a wide range of services and if there were too many, they would not survive. Only a few locations are suitable for growing into large settlements. **2. A** residential **B** recreational **C** commercial **D** industrial **E** residential **F** recreational **G** recreational **H** waste land **I** transport **J** recreational **K** recreational **L** residential **M** recreational

Models

In large towns and cities, it is often possible to see a clear pattern of land use which reflects the development of the settlement. The distribution of urban land use can be explained using **models of urban land use** such as those shown below.

- ■ Central business district (CBD)
- ■ 19th century housing and industry
- □ Older high class housing
- ▨ Modern housing
- ▨ Modern industrial estate

Concentric rings model where the city has grown outwards from its old centre.

Sector model where the city has grown around transport lines such as canals or railway lines.

Multiple nuclei model where the city has grown around more than one centre (or CBD).

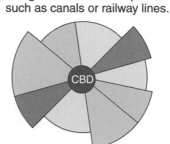

local business district

commercial district

Check this out using an Ordnance Survey map. Cities such as Glasgow, Newcastle, Leeds and Manchester are good examples. You should be able to decide which of the models above can be used to explain the city's development.

Cities in the Economically Less Developed Countries are rather different from those in the Economically More Developed Countries. A model of an ELDC might look like this:

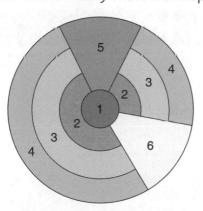

1. **Central Business District**
   This is the main commercial area of shops and offices. It is very similar to any city in Economically More Developed countries.

2. **Medium cost housing** This contains older housing, built when the city was smaller. The wealthiest inhabitants have now moved away from this area.

3. **Low cost housing** This is made up of older squatter (shanty) housing which has been improved over the years, or government built, low cost housing for workers.

4. **Shanty towns** These areas are built by new settlers, using basic materials such as corrugated iron, wood and other scrap materials. There are often no basic amenities such as water supply, sanitation or electricity.

5. **High cost housing** This area contains high cost apartments or town houses in the most attractive areas, often close to main roads and the CBD.

6. **Industrial areas** Industrial areas have developed along main roads, railways or rivers.

The biggest single difference is that most cities in Economically Less Developed Countries suffer from mass migration of people from rural areas. The vast majority of these people are very poor and cannot afford to buy or rent houses. They often live in houses or shacks they have built themselves (**shanties**) which are located at the edge of the city.

Site, situation and function

Where settlements are found 1

The initial location of a settlement is influenced by its **site**.

The site of a settlement is the actual area of land on which it is built.

The original site of a settlement may have been chosen for a number of reasons, for example:

- On a hill for defence
- On dry land above a marshy area
- Beside a sheltered cove for ships to load and unload
- Beside clean water for drinking, cooking, etc.
- Beside a shallow crossing place on a river.

Quick Test

Describe the original site of the settlement where you live.

Answers An example answer might be: The town was established beside the River Forth where it was easy for people and cattle to cross. In time a bridge was built and traders settled to cater for travellers. The river provided water for drinking and washing, and in due course a mill was built, using water power.

Where settlements are found 2

Even if the site of a settlement is excellent, it is unlikely to grow into a large town or city unless it has a good **situation**.

The situation of a settlement relates to its whereabouts. This is influenced by good routeways, access to a harbour or a raw material such as coal.

Quick Test

Describe the situation of the settlement where you live.

Answers An example answer might be: The town grew as trade increased. Roads followed the river valley and the building of a bridge meant the town became a crossroads. In later years the river was deepened and ships could be berthed. In the 19th century, coal deposits were discovered nearby and the population grew rapidly, as did the port.

The function of settlements

Each settlement has a definite character of its own. This is largely due to its **function**.

Examples of the functions of settlements include:

- a market town (e.g. Lanark)
- a holiday resort (e.g. Blackpool)
- a port (e.g. Dover)
- a mining town (e.g. Motherwell)
- an administrative centre (e.g. Edinburgh)
- a manufacturing town (e.g. Kirkcaldy)
- a centre of learning (e.g. St Andrews)

Top Tip

Know what the terms 'Comprehensive Redevelopment' and 'Urban Regeneration' mean and be able to apply each of these to a city you have studied.

Some settlements have several functions. St Andrews, for example, is a holiday resort as well as a centre of learning.

Think of a further example of each type of settlement listed above, one in Scotland and one in the rest of the United Kingdom.

Comprehensive Redevelopment is usually used to describe the way in which large areas of the inner city were demolished, often in the 1960s and 1970s, to be replaced by high rise flats. It was usually coupled with the removal of many of the original population to new towns or housing estates on the edge of the city.

Urban Regeneration describes the changes in a city over the last 50 years. It includes redevelopment of the inner city as well as modernisation of the CBD, the introduction of traffic management schemes, the development of new industry, often nearer the city edge, and the refurbishment of older properties, both residential and commercial.

Quick Test

What is the function of the settlement your school is in?

Answers Every town has a different range of functions. Your school could be in a market town, a former mill town or near a harbour where fishing was once important.

Services and spheres of influence

What you find in settlements

Settlements provide **services** to those who live or work in them, or travel to them for these services.

The bigger the settlement, the more services it usually provides. A village, for example, might only have a small shop and post office, a telephone box, a public house and an hourly bus service to the nearest town. These can be described as **low order services**.

A large town, on the other hand, will have many services. Not only will it have many more shops and bus services, it will also have specialised services like department stores, bookshops and cinemas. Services such as these are called **high order services**. They rely on custom from a large population in order to be profitable, so they are only found in towns or cities. You should be able to think of examples of low order and high order services in your local area.

Quick Test

Decide which of the following services are high, medium or low order services. Present your answers in a table.

- Petrol station • Library • Café • Supermarket • Hairdressing salon • Electrical store
- Post office • Braehead/The Gyle/ Metro Centre shopping centres • Concert hall (e.g. SECC)
- Police station • Tanning studio • CD/DVD store • Furniture warehouse • Garden centre

Answers (You might want to discuss some of the choices!)

LOW ORDER	MEDIUM ORDER	HIGH ORDER
Café	Library	Electrical store
Petrol station	Supermarket	Shopping centre
	Hairdressing salon	Concert hall
	Post office	CD/DVD store
	Police station	Furniture warehouse
	Tanning studio	Garden centre

Spheres of influence

Services attract people from the surrounding area who want to use them. This area is called the **sphere of influence** of that particular service.

A low order service, such as a newsagent, will only attract people from close by, so it has a small sphere of influence.

A high order service will have a larger sphere of influence, as people are prepared to travel farther to use it. Hampden Park and Murrayfield are good examples of high order services, as they attract football and rugby fans from all over Scotland to watch important international matches.

Comparing services

The diagram below shows the spheres of influence of two different services: a village store and a large supermarket in a town.

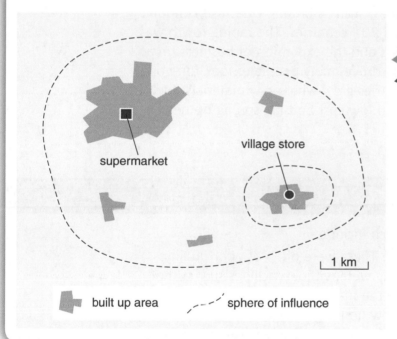

Top Tip

Sometimes high order services are found in the most unlikely places. A small rural town, for example, might have services such as saddlers or auction markets which are usually more typical of large settlements. A good example is Wigtown, in Dumfries and Galloway, which has established a reputation for bookshops – people travel large distances to visit them.

Exam-style Question

What are the advantages and disadvantages of building large retail parks (e.g. Braehead or the Gyle) on the edge of the urban area?

The problems of urban growth

Too many people?

The population of the world continues to grow, and more and more people decide to live in urban areas, as this is where opportunities such as jobs are to be found.

As urban areas grow larger, problems are often created; for example, pollution, congestion, overcrowding and stretching of resources like water supply or sewage disposal. Cities in **Economically More Developed Countries** (**EMDCs**) and cities in **Economically Less Developed Countries** (**ELDCs**) have different problems.

Urban growth 1

In Economically More Developed Countries, such as Britain, cities like London grew in size through most of the 19th and 20th centuries. The rapid growth was largely due to developing industries, ports and the expansion of business. In a similar way to ELDCs today, rural poverty drove many people to look for work in the rapidly growing urban areas. Industries grew where raw materials, fuel supply and markets were to be found, and low cost housing sprung up near these factories.

Urban problems 1

This has often led to urban problems which include:

- Poor quality housing (e.g. **tenements**). These were often of poor quality, totally unsuited to modern living. Many were without facilities such as hot water, toilets and sufficient rooms and the majority were demolished in the 1950s and 1960s to make room for new housing or high rise flats.

- **Urban sprawl** with **suburban growth**. With the growth of tram and rail routes, and the later development of car transport, routes leading into the city were attractive to housing developers, who built extensively throughout the first half of the 20th century. Housing estates followed the main roads and railways and large areas of land were suburbanised.

- Greater travel distance to work. The growth of rail and road transport and an increased desire for many people to live in the countryside or rural settlements, have meant that many commuters spend a lot of time and money travelling to and from work every day.

- Lack of car parking space, inadequate public transport and the building of urban motorways and ring roads. Most city centres were developed over 100 years ago, before the advent of the motor car. City streets are totally unsuited to carry the volume of modern traffic, so large car parks are necessary for those who do drive to the city. Through traffic has had to be routed around

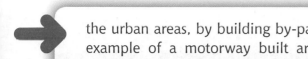

the urban areas, by building by-passes and motorways. The M8 in Glasgow is a good example of a motorway built around the CBD. Its construction required the demolition of housing and industry; the M74 extension will cause similar problems.

- **Traffic congestion** and pollution. The huge increase in car ownership has made travel to work a daily habit for many workers and this causes congestion and pollution in both rural and urban areas.

- Loss of jobs in the traditional industries and docks, leading to urban dereliction and waste ground. Many of the industries which grew in inner city areas became inefficient as technologies developed. Demand for their products fell (e.g. steam engines and metal products) or could be provided more cheaply by countries with a lower paid workforce (e.g. shipbuilding) and most closed. This led to unemployment, poverty and those who looked for new work often had to travel to industrial areas on the edge of the city or new developments such as New Towns.

Top Tip
You should understand how each of the problems mentioned has come about and also how it has affected one city you have studied.

Urban growth 2

In Economically Less Developed Countries, such as India or Brazil, the sudden growth in city size has been particularly obvious in the last fifty years. Rapidly increasing rural population has led to **migration** from the countryside to the towns and cities.

Urban problems 2

Top Tip
You should be able to describe some of the problems facing people living in a city in an Economically Less Developed Country.

The problems of cities in the Economically Less Developed Countries are therefore rather different to those in Economically More Developed Countries:

- Overcrowding, with many people living in sub-standard housing – these may be shanty towns, often built on the very edge of the city;
- Insufficient work for those who live in the city;
- Poverty and disease;
- Lack of services (e.g. medicine) and food for the ever-expanding population;
- Traffic congestion and uncontrolled air pollution.

In the early 20th century, the world's largest cities were all in the Economically More Developed Countries; now the vast majority are in the Economically Less Developed Countries.

Quick Test

Give three reasons why cities in ELDCs are growing so quickly.

Answers • Rural poverty forcing people to move to urban areas • Jobs and services available in cities • High rates of population growth.

Urban change

Tackling urban problems

The problems of urban growth described in the previous pages have been tackled in many ways, some more successful than others.

In Economically More Developed Countries

Cities in Economically More Developed Countries (such as Britain) have:

- Knocked down many of the old buildings near the city centres and replaced them with multi-storey flats
- Built large housing estates at the edge of cities to accommodate people who used to live in old housing near the city centre
- Built **New Towns** some distance from the city with the attraction of new jobs and more pleasant surroundings
- Renovated many of the old buildings which were structurally sound but lacked modern facilities
- Created new areas of housing on the sites of old factories, often near the city centre or docks
- Attracted new industries into the areas cleared of old industries. These can be called **brownfield sites**.

You should be aware of the advantages and disadvantages of each of the solutions above, in terms of:

- The **social impact** on family, friends and travel
  Many families were relocated in new areas, either on the edge of the city or in new housing developments. Neighbours may have been moved to other areas, so friendships were lost. People moving to New Towns such as East Kilbride or Cumbernauld found themselves far from friends and relatives. New housing developments had few services such as shops, cinemas, pubs or social centres, so people felt they were deprived. Living in a healthier area had a positive health benefit, however, with less disease and improving quality of life.

- The **economic effect** on employment and shopping
  Many people would have worked close to their homes in the inner city; after moving, they had much farther to travel to work, to shop and for entertainment. This was expensive, and as many of those relocated were working class families, money was often limited.

- The **environmental impact** on landscapes
  The areas which had to be redeveloped were environmental eyesores, often with poorly built, cheap housing close to heavily polluting factories; new houses were better equipped, more attractive, in healthier areas and away from noise, smoke and other dangers. Many factories were knocked down, replaced by newer, light industries.

In Economically Less Developed Countries

Cities in Economically Less Developed Countries have very different problems:

- The quality of life for those living in shanty towns needs to be improved. Better education and medical facilities, and services such as a clean water supply, electricity, public transport and sewerage, need to be provided. Some city authorities have encouraged settlers to build better houses by providing cheap materials or loans; others have built low cost, basic houses or flats for those formerly living in shacks.

- People who have moved to the city need to find employment, so industries should be encouraged to locate factories close to the edge of towns. People with an income are then able to improve their houses and standard of living.

- Some authorities have taken tough action with people living in shanty towns, even to the extent of bulldozing down their homes and forcing them to leave the area. People living rough on the streets can be arrested and removed from the city.

- Many Economically Less Developed Countries are trying to slow down the rate of migration to the cities. In Brazil, for example, money is being invested in the rural areas in the north-east of the country to encourage people to stay put rather than migrate to the cities. Money is spent improving education and medical facilities, and help is given to farmers to irrigate their land and improve farming methods to reduce poverty and hunger which force people to leave the countryside.

Top Tip

You should be able to describe how one city or country you have studied has tried to overcome some of its problems.

Quick Test

Give four examples of the ways cities in ELDCs have tried to overcome their problems.

Answers • Self-help schemes to encourage people to improve their houses in shanty towns • Persuading industries to locate in cities to provide employment and wealth • Demolishing shanty towns • Encouraging people to stay in rural areas rather than migrate to cities.

Farming systems

The farming process

Any farm can be thought of as a **system**, with **inputs**, **processes** and **outputs**. This is true of farms in any part of the world.

A cycle

Farming can also be thought of as a cycle, which usually takes one year to complete.

The diagram below shows a very simple farming cycle:

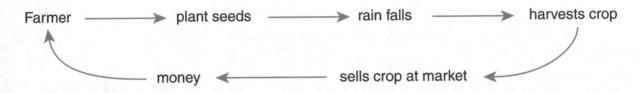

Farmer ⟶ plant seeds ⟶ rain falls ⟶ harvests crop

money ⟵ sells crop at market ⟵

Quick Test

Can you identify two inputs and two outputs in the farming system above?

Answers Inputs are: seeds, rainfall and farmer's decision. Outputs are: crops, money.

What goes in ...

In real life, farming systems are usually much more complicated. There are many different kinds of **inputs** which might include:

Types of input	Examples
Physical inputs	soil, rainfall, sunshine, altitude
Human inputs: economic technological social political	profitability, rent, transport costs machinery, fertilisers, pesticides labour, demand government subsidies

What goes on ...

There are also many different **processes** taking place on a farm. Just which processes depend on the type of farm, but examples might include planting, grazing, milking, ploughing, spraying insecticide or harvesting.

What comes out ...

In the same way, the **outputs** from a farm will vary. They might include the crop grown on the farm, milk, calves, wool, grapes, cotton and, if the farm is profitable, money. In poorer countries, where many farmers only grow enough food for themselves and their families, money is seldom an output. Farms like this are called **subsistence** farms.

Options

It is important to realise that farmers are **decision makers**. The decisions they make will affect their farms' profitability. A British farmer will have many more choices available to him or her than a farmer in a country such as Ethiopia. You should be aware of the kinds of decision a farmer has to make, and the consequences of such decisions. These could include:

- Selecting from a number of crops to grow. This decision will be based on profitability, likely market price, the farmer's land and the likelihood of a good harvest.
- Whether to invest in new machines or employ more (or fewer) workers.
- Borrowing money from the bank.
- Diversifying into other interests such as holiday homes or renting fields out for pop concerts.
- Leaving fields uncultivated for a number of years.
- Deciding to turn the farm into an organic one.

Changing farms

The modern farmer in Britain has many issues to consider. Many traditional crops and farming practices are uneconomic because of changing lifestyles, increasing industrialisation and a global market. **Diversification** is a route many farmers are trying. Examples include:

- Converting old farm cottages and barns into holiday homes
- Recreational activities; for example building golf ranges
- Allowing wind farms or mobile phone masts to be erected on their land
- Opening farm shops
- Rearing rare species for showing or for specialised food markets.

Top Tip
Be aware that many of the changes facing farmers have both advantages and disadvantages.

Different types of farm

Different farms

Although you may study farming processes on different types of farms, the only types of farms you will have to know about for the exam are those found in Britain. These are **arable** and **pastoral**. Some farms combine arable and pastoral; these are **mixed farms**.

Arable farms

Arable farms produce crops. You should have some idea of the types of inputs and outputs for an arable farm, along with some of the processes which take place on the farm. The inputs might include seeds, fertilisers, capital investment, machinery (e.g. tractors, harvesters, ploughs), water for irrigation, labour, storage facilities, soil improvers. The processes will include thinning, harvesting, sorting, transporting the product. The outputs are the crop produced, any waste products (e.g. straw, potato shaws) and profit.

In the exam, you may be faced with a question about a type of arable farming in an area you have never studied or know little about. Don't panic! You will not need to know any details about the crops produced or the local area. If you do need to know anything, then the relevant information will be given to you in the question.

What you know about arable farming from your coursework will let you answer the question. This would include such things as:

- Why are arable farms found in drier, warmer areas?
  (Warmer, drier soils are more suitable for crops, more sunshine/warmth for ripening and harvesting)
- What have been the main changes in farming processes in recent years?
  (This could include the removal of fences/hedges between fields, greater use of machinery, a reduction in the number of farm workers, larger farms produced by **amalgamation**, the growth in **organic farming** methods, the introduction of **genetically modified** products, or the importance of European Union subsidies and policies such as quotas and set-aside policies.)

Pastoral farms

Pastoral (or **livestock**) farms rear animals – for their meat and for their produce (e.g. milk or wool). Once again, you should know something about the inputs, processes and outputs from pastoral farms. Inputs might include young animals for fattening or breeding, fodder, winter shelter and pesticides. The processes will vary depending on the livestock kept but might include grazing, milking, shearing, fattening and the outputs could be wool, leather, milk or animals for meat.

You should be able to answer:

- Why are pastoral farms normally found in higher, wetter or colder areas than arable farms?
  (Grass grows well in these conditions to produce summer food and winter fodder such as silage or hay)
- What are the main problems facing today's pastoral farmers?
  (Reduction in demand as meat consumption is falling, quotas imposed by government/ European Union, the impact of scares such as BSE and Foot and Mouth disease, cheap imports from abroad resulting in falling prices)
- What changes have come about in recent years?
  (Changing public tastes for animal products, the introduction of bioengineering techniques, tighter restrictions on production methods)

Some issues facing farmers today

There are also a number of issues which affect all farmers. Make sure you can answer:

- Why is there concern about **intensive** farming methods?
  Intensive livestock production can raise welfare issues; intensive crop production can encourage the deterioration of soils and overuse of chemicals
- What effects have recent changes had on the environment (e.g. the greater use of fertilisers and pesticides, the introduction of **set-aside** land)?
  More fertilisers and pesticides can damage wildlife such as birds and animals or pollute water supplies. Set-aside land, where farmers are paid not to use fields for production, is a waste of a resource which many people find unacceptable in a world where starvation is common.
- Increasing mechanisation has led to larger fields, with hedgerows being removed. This has led to a reduction in wildlife habitat for many birds, animals and wild flowers.

Top Tip
You should know about both advantages and disadvantages of agricultural change in the last fifty years. Think of the farmer, the people living in the countryside and the consumer.

Exam-style Question

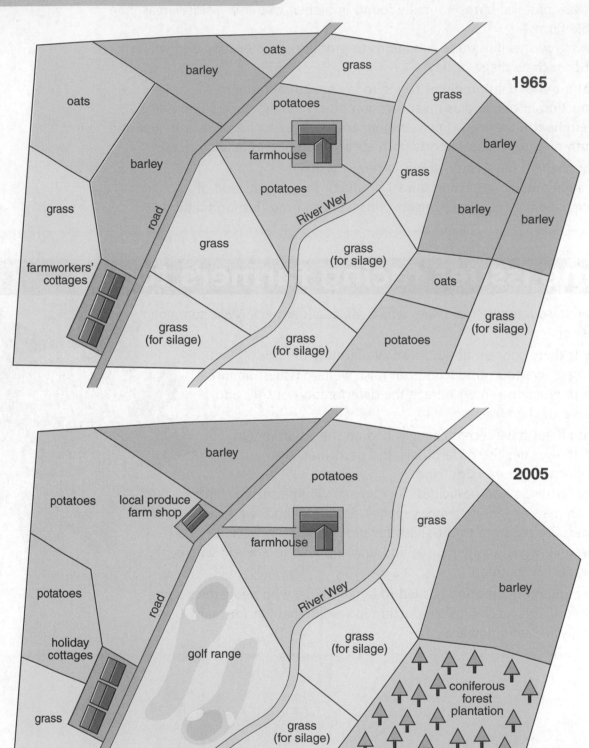

(i) Give reasons for the changes in land uses on Holm Farm between 1965 and 2005.

(ii) A group of students is carrying out fieldwork on Holm Farm to record the relationship between land use and relief and drainage. Suggest two field techniques they could use to gather information, giving reasons for your choices.

(iii) Many farming areas in Britain are suffering from rural depopulation. Describe why this is the case and explain the problems this may cause to the local area.

Classification

Different types of industry

Industry can be classified as **primary**, **secondary** or **tertiary**.

Secondary (manufacturing) industry can also be described as **light** or **heavy industry**. You should know at least three examples of each.

Quick Test

1. Give three examples of each of the above

2. Complete a table to show which of the following industries are light or heavy industries.
   - Shipbuilding
   - Oil rig construction
   - Aero engineering
   - Biotechnology
   - Manufacture of computer equipment
   - Manufacture of mobile phones
   - Making industrial chemicals
   - Manufacture of televisions

Light	Heavy
Computer manufacture	Shipbuilding
Mobile phone manufacture	Oil rig construction
Biotechnology	Aero engineering
Manufacture of TVs	Industrial chemicals

Answers 1. Answers could include: **Primary:** farming, forestry, fishing
Secondary: brick-making, car manufacture, making paper
Tertiary: nursing, vet, lorry driving

What influences where industries are found

Industrial location is dependent on a number of factors. These include:

PHYSICAL FACTORS:
raw materials
power
site
transport
climate

HUMAN FACTORS:
labour requirements
markets
capital
government policies

You should know which factors were important to the choice of location for a number of industries.

Examples

Motorola (circuits, mobile phones) in East Kilbride:
Close to good transport links for components and finished products (close to M74, M8, Glasgow and Prestwick Airports); source of skilled workers (large town, universities in west of Scotland); government incentives for factory building; available land on technology park.

Torness Nuclear Power Station at Dunbar:
Close to main road (A1), main railway and sea for transport of nuclear fuel and waste; near population centres of central Scotland; site on flat land near coast (safety); government support.

Caledonian Paper at Irvine:
New Town site with available labour force; flat land for large factory; Government incentives for new factories; close to port and main roads for inward transport of raw materials and outward transport of finished papers.

BP Grangemouth Oil Refinery:
Very large site on flat land on estuary of River Forth; close to sea for tanker unloading/pipeline terminal from North Sea oil resources; large workforce in area of high unemployment; government incentives; close to large demand for fuels and products in central Scotland.

Industry is always changing

Industry undergoes change through time.

- Many industries have closed down because they were inefficient, too far from modern markets, or their product was no longer in demand. These are often described as **traditional industries** or sunset industries. Examples include steel making and shipbuilding.

- Other industries **relocated**, because their source of raw materials changed and/or their markets changed. Examples include breweries and integrated steelworks.

- Some industries have undergone **rehabilitiation** to meet the requirements of modern demand. This has involved modernised buildings and/or updating machinery. Newspaper publishing is a good example.

- The most recent development in industry has been the dramatic growth of **high technology industries** which specialise in microelectronics or biotechnology. Sometimes these are called **sunrise industries**. Examples include making microprocessors for computers and companies involved in genetic research.

Top Tip
Many industries stay in particular locations long after the reasons they located there in the first place have disappeared. This is called **industrial inertia**. This is why there is still steel manufacturing in south Wales, or a textile industry in Yorkshire.

Industrial change

Changing industry

The location of industry and the importance of different industries change over time.

Primary industry

As raw materials become exhausted or cheaper sources become available, mines close down. Over-fishing in certain areas and changes in the demand for farm produce also cause changes in the location of primary industry.

You should be able to describe the decline of an industry you have studied in a particular area. Two good examples are coal mining in central Scotland and fishing in north-east Scotland.

The closure of coal mines meant local communities experienced a number of changes, some negative, some positive.

Negative effects	Positive effects
• High unemployment • Reduction in wealth • Closure of many support services such as shops • Families living on benefit	• Closure of unsightly mines • Safer environment • Less pollution from associated factories • New industry attracted into area

The decline of the fishing industry had a similar effect on the coastal towns and villages of north-east Scotland.

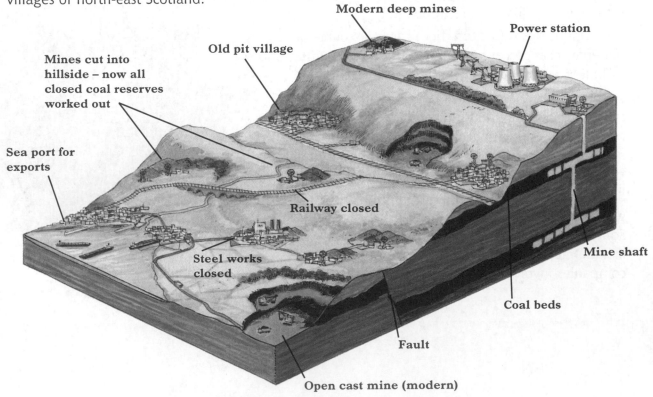

Secondary industry

As the source of raw materials changes, the demand for products changes, or factories become out-dated, the location of manufacturing industry also changes.

You should be able to describe why a particular industry is no longer important in a particular area. Examples could include shipbuilding on the Clyde or steel making in Lanarkshire or the Ruhr (Germany).

The impact on local communities is very similar to the closure of coal mines described on the previous page.

Tertiary industry

The more developed a country becomes, the more people work in service industries.

You should be able to explain the growth of a service industry over the past twenty years. Examples could include:

- tourism in Scotland
- telesales
- the opening of the Channel Tunnel.

Quick Test

Write a short paragraph to explain why **one** of the service industries shown above has grown rapidly over the last decade.

Answers Your answer might include reference to some of the following: higher income meaning more money to spend on pleasure; the growth of information technology; more people travelling on holiday, including second holidays; more competition in the market place.

Industrial decline

When traditional industries close down, they often have a severe impact on the local area. You should be aware of the **social, economic** and **environmental effects** of industrial decline on an area you have studied. You might be able to think of benefits too, such as reduced pollution or fewer ugly scars on the landscape.

The social effects of industrial decline describe the impact on families and individuals. Reduced income and spending power can cause families to suffer stress and marriages can fail. In severe cases there may even be an increase in crime as some people resort to extreme measures to compensate for lack of money.

Top Tip
Remember new industries such as electronics can also close down and have a negative impact on the local community.

The economic effects describe the financial impact on individuals and the community. With less money around, families have to reduce their spending by taking fewer holidays and buying fewer luxuries. For many there is the possibility of getting into debt as paying bills becomes increasingly difficult. Local communities suffer too; shops may have to close and services such as pubs and clubs may have their profits reduced.

The environmental effects can be both positive and negative. On one hand, there might be less pollution from factories and ugly buildings may be knocked down, but sometimes old works lie empty and become dangerous eyesores.

Population structure in rural areas

Population change

The population of an area depends on:

- The rate of natural increase. In other words, how many babies are born (**birth rate**) and how many people are dying (**death rate**).
- The rate of **migration** into or out of an area.

In Economically More Developed Countries

In Economically More Developed Countries, rural areas often experience:

- A declining population, as people move from a relatively poor area to wealthier areas with better job prospects. This is called **rural depopulation**.
- An ageing population, as most of the people who move away are young (ambitious) people. These are the people who are more likely to have children.
- A reduction in services such as schools, health care, public transport and shops as there are fewer people left to support them.

In Economically Less Developed Countries

In Economically Less Developed Countries, the problems are similar but much more noticeable. Migration from rural areas is often very great because of:

- Rapid population growth in less developed rural areas means too many mouths to feed for the amount of food available.
- Increasing poverty and starvation or malnutrition.
- The chance of work and better living conditions in the cities.
- The quality of health and education services is often poor.

Rural regeneration

It is worth noting that, in some areas, people are being attracted back to rural areas. In Scotland, for example, towns like Aviemore and Invergordon have encouraged people to move there because of the development of manufacturing industries or tourism. This is often heavily supported by government grants in an attempt to reduce rural depopulation.

Top Tip
Be aware that population movement can provide opportunities as well as problems. Incoming workers can often find employment in lower paid jobs that are traditionally hard to fill. Skills shortages can also be overcome.

Quick Test

Why is there a decline in the number of services in rural areas such as the Scottish Highlands or rural Wales?

Answers As people move away from these areas, there are fewer people to support local shops, petrol stations, health centres, bus services, police stations, schools, hospitals, etc

Migration: push and pull factors

Why people leave the countryside 1

Rural migration is usually explained by

PUSH FACTORS   which force people to move away, and

PULL FACTORS   which attract people to urban areas.

Push factors are things which make people feel dissatisfied with where they live and encourage them to think about leaving. They might include remoteness, low wages, poor employment opportunities, being away from 'the buzz' or limited services such as shops or entertainment. In some parts of the world they might be fear of violence or discrimination, extreme poverty, starvation or even war.

Pull factors are things which people see as attractive about an area. There might be better employment or education prospects, higher wages, the attraction of city life or the chance to make a way in the world. In some countries it could include security or an improved standard of living.

You should be able to describe the push and pull factors which operate in Economically More Developed Countries as well as in Economically Less Developed Countries.

Why people leave the countryside 2

People who move are called **migrants**. There may be **social**, **economic** or **political reasons** for moving away from rural areas.

In an Economically More Developed Country, **social reasons** are likely to relate to poor opportunities and distance to services such as shops, entertainment or health facilities.

In an Economically Less Developed Country, **social reasons** are more likely to be over issues such as health, welfare and poverty. Families often find themselves under huge pressure if there is insufficient food or income to maintain a reasonable quality of life. There may be illness or disease relating to poor living conditions and hospitals and schools may be few and far between.

Economic reasons relate to the poverty found in rural areas. In an Economically More Developed Country, incomes are often lower than in urban areas so there is a degree of rural poverty which encourages migration to more prosperous centres. This is sometimes linked to education prospects – universities and colleges are seen as the opportunity for a better job and these tend to be found in cities.

Top Tip

You should know in some detail about the effects of migration on:
- A rural area in an Economically More Developed Country (e.g. rural Wales)
- An Economically Less Developed Country city (e.g. Mexico City or Calcutta)

In an Economically Less Developed country, the problem is similar but far more extreme. Many rural areas have virtually no job prospects and poverty is widespread. Linked to increasing population, the situation gets worse rather than better, so there is a strong reason behind the large-scale migration to the cities. Poverty means that, in times of famine, families are unable to survive without moving to urban areas where food supplies are concentrated.

Political reasons for migration are normally caused by governments or ethnic groupings.

In an Economically More Developed Country, these do not normally feature, although historical happenings such as the Highland Clearances, during which thousands of people living in the Highlands of Scotland in the 19th century were forced off their land, are a not too distant event.

In an Economically Less Developed Country, this is a more common occurrence. Some governments discriminate against certain elements of the population and fear of torture, imprisonment or even death forces many to seek a new life in another country. When people are forced to move and leave everything behind, they are called **refugees**.

Cities in Economically Less Developed Countries are growing much faster than cities in the developed world. This is a result of very powerful Push and Pull factors caused by a combination of extreme rural poverty and rapidly growing population in many parts of the world.

Quick Test

Why were most of the largest cities in the world fifty years ago in the Economically More Developed world, whereas today the largest cities are in Economically Less Developed Countries?

Answers Through the 19th and first half of the 20th century, cities in the EMDCs grew as industry and commerce developed. In the last fifty years, the combination of rural poverty and rapid population growth in ELDCs has meant that people have flocked to the cities in search of a better life. In EMDCs, population growth has slowed drastically, and family wealth has, on the whole, improved.

Distribution of world population

Population distribution

The world's population is very unevenly distributed. In order to explain this, it is often better to think about **population density** rather than the total population of a country or region. Population density is described as the number of people per square kilometre.

You should be able to look at a map showing world population density and explain why some areas are densely populated while others are not. This could be due to:

Environmental reasons

In areas where the environment makes settlement difficult or dangerous, population densities tend to be low. Examples include:

- Hot deserts
- Tundra and cold deserts
- Rainforest
- Mountainous areas or places with rough terrain
- Areas prone to frequent flooding.

The world's most populated places are often in areas with a climate that is not too extreme (temperate).

Economic reasons

In the past, populations developed in areas where there was a reason for settling there. Examples include:

- Fertile farmland
- The availability of natural resources such as timber, fuel, metal deposits
- Near transport routes such as ports, rivers, easy routeways.

Political reasons (e.g. war or resettlement)

Populations are also concentrated in areas determined by political decisions. Examples include:

- Settlements planned or ordered by governments (New Towns in Britain; Brazilia, the capital of Brazil; the densely populated state of Israel are examples)
- Wars can also have an influence on where people settle. The movement of Jews after the Second World War and the results of ethnic wars in Rwanda in the 1990s are two examples. The black population of North America and the Caribbean is another – a result of the slave trade of the 18th century.

Counting people

Population details are gathered in each country by a **census**. In many countries this is not very accurate. This may be because the people completing the details may not be very literate, or some people in an area may be so spread out that they can be missed. Sometimes it is possible to avoid paying taxes or fighting in a war by not being included in the census.

Top Tip
A census is normally carried out every ten years. You could argue that it is out of date before it is published, but it still provides valuable information for governments.

Showing population change

As explained on page 76, the rate at which the population changes depends on the relationship between **birth** and **death rates**. If the birth rate is greater than the death rate, the population increases. This can be shown on a graph, which, over a period of time, can be divided into three phases.

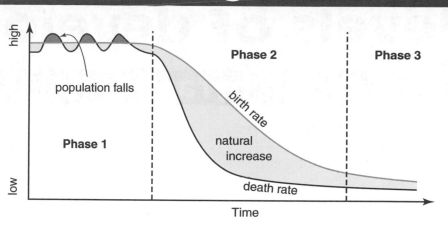

You should be able to state what is happening to the total population of a country in each of the three phases, and explain why birth and death rates change over time.

In phase 1, birth rates are high because:

- Many people choose not to practise birth control.
- With a high infant mortality rate, it makes sense for families to be large to ensure a number of children survive to maturity.
- Life expectancy is often low so there is a need for a large family support as parents reach old age and are less able to work.
- Death rates are high because there is a lack of health care and disease is common.
- The death rate can occasionally rise above the birth rate because of epidemics or war.
- Countries in this phase are experiencing a steady population with fluctuations caused by disease or natural disasters (e.g. famine).
- Examples – Bhutan, Papua New Guinea.

In phase 2, the death rate falls quickly and the birth rate remains high because:

- Improved medicines and health care result in improved life expectancy and fewer deaths through disease.
- The tradition of having large families lingers for some time.
- Birth control measures may take time to be accepted.
- Countries in this phase are experiencing rapid population growth, sometimes called a population explosion.
- Examples – Mali, Bangladesh.

In phase 3, both the birth rate and the death rate are low because:

- Life expectancy continues to improve.
- More people live until old age.
- The adoption of birth control is widespread and accepted.
- The economic benefits of smaller families is seen.
- Countries in this phase are experiencing slow or no population growth.
- Examples – Switzerland, USA.

Quick Test

Give three reasons why officials carrying out the census in a ELDC might have difficulties doing so accurately.

Answers Remote population (difficult to trace); people avoiding census; poor literacy; (people have difficulty completing forms).

Population pyramids & levels of development

Population pyramids

Population pyramids are a useful way of comparing the populations of different countries.

You should be able to recognise the shape of a pyramid which represents

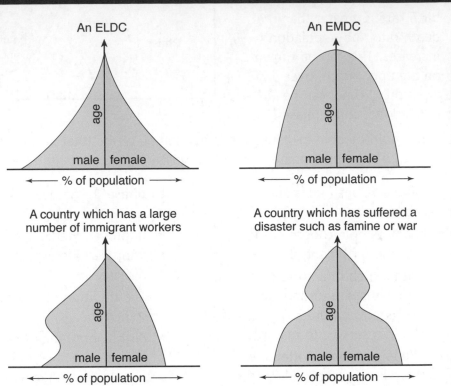

Indicators of development

It is also possible to compare Economically Less Developed Countries and Economically More Developed Countries by looking at a number of social and economic factors.

The table below shows the differences you would expect to find in some of these, but remember there are many others which could be used:

Measure of development	ELDC	EMDC
% literacy	low	high
% working on farms	high	low
% population under 15 years	high	low
% population over 64 years	low	high
Hospital beds per 1000 people	low	high
Car ownership	low	high
Gross National Product (GNP)	low	high
Population growth	fast	slow
Life expectancy	low	high

Top Tip

There are many other indicators of development, including:
- infant mortality
- % of population with access to clean water
- average income
- average size of family
- televisions per 1000 people
- average calorie consumption per day

Exam-style Questions

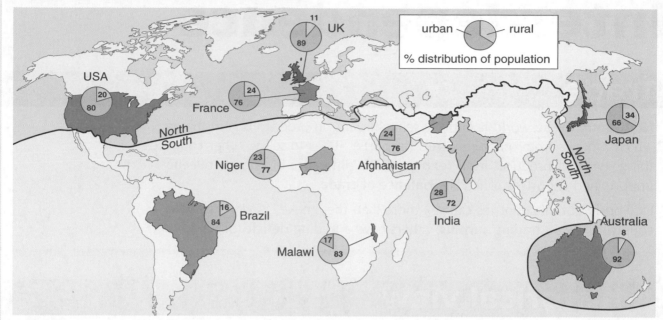

1 Refer to the world map: Urban/rural distribution of population in selected countries (2004).

(i) Describe the pattern of urban/rural population shown in the world map.

(ii) Explain the differences in pattern between the North and the South.

2 Problems facing those moving to urban areas in the Economically Less Developed Countries:

- Lack of employment
- Shortage of housing
- Overcrowding
- Poor sanitation and health
- High cost of living

Explain why many people continue to move from rural areas to cities in the Economically Less Developed Countries.

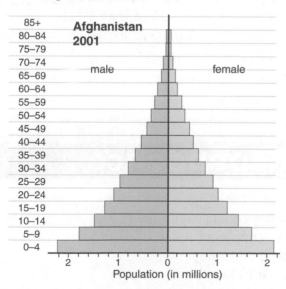
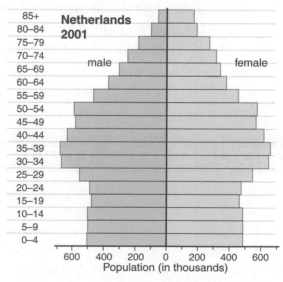

3 Give reasons for the differences between the population structures of Afghanistan and The Netherlands.

For each of the countries above, explain the problems that face the governments in providing services for the populations.

Trade and interdependence

Trade

No country in the world is entirely self-sufficient. In order to buy in things which it needs from other countries (**imports**), it has to sell to other countries things which it produces itself (**exports**). The difference in value between imports and exports is called the **balance of trade**.

Obviously not all countries can sell more than they make, so while some countries have a trading **surplus**, others have a trading **deficit** (or loss).

A historical view

The countries which developed earliest (such as those in Western Europe or North America) did so because they were able to turn raw materials into manufactured products which were then sold to produce wealth. Many of these raw materials came from their **colonies**.

A historical inequality?

Many Economically Less Developed Countries, which were often former colonies, are still the main producers of raw materials which they export to the Economically More Developed Countries. Unfortunately for them, the price they are paid for the raw material is only a fraction of the value of the item it is made into. Often they have to import the finished item from the Economically More Developed world. In many cases the gap between rich and poor countries continues to increase.

What makes things even worse is that the price paid for the raw material is often controlled by the Economically More Developed Countries.

Trapped in a trading dependence

The odds are stacked against Economically Less Developed Countries in the trading world. As explained above, many are tied into dependence with the Economically More Developed Countries because of the crippling burden of debt they have. Because so much of their income is devoted to paying these debts, or interest on them, there is little or no money left to invest in building factories which could process the raw materials which would command a higher price on the world's markets. In many cases, the Economically More Developed Countries see the poorer countries as a source of cheap products which they themselves can process to make money. Products such as timber, coffee, metals and cotton are examples.

It is only when Economically More Developed Countries are prepared to reduce the debt of Economically Less Developed Countries or help them invest in manufacturing plants that this inequality will start to change.

You should be able to explain how a country can become trapped in its dependence on another country, and describe some of the ways in which this dependence can be broken.

Many of the Economically Less Developed Countries make most of their money by selling exports to the Economically More Developed Countries of the world. Very often, a large part of their income from trade is from the sale of only one raw material. This makes them very vulnerable to variations in the price paid for that item. Zambia, for example, receives over 90% of its export income from the sale of copper.

You should be able to describe the effects that a drop or rise in the price of copper would have on Zambia. This would include employment, earnings, standard of living, investment in industry, education and health, and pressure on the government.

Beating the inequality

The price of manufactured goods has risen far faster than the prices paid for raw materials. As suggested on page 84, Economically Less Developed Countries import mainly manufactured goods, so the money they make from their exports doesn't go as far as it once did. One way of overcoming this problem is to enter into a **countertrade** agreement. Instead of receiving money for exports, the developing country receives goods. More and more countries are trading in this way.

Another way of getting more value for exports is to encourage **Fairtrade** agreements. This means that companies, often supported by charities, buy products at a fair price from the producers and this helps to reduce poverty and improve the standard of living. Produce such as coffee, tea and fruit are increasingly sold in this way.

Top Tip
A Transnational (or Multinational) corporation is a large company with offices and/or factories in several different countries. Volkswagen and ICI are good examples.

Exam-style Question

Malawi – fact file

Development indicators

- GDP per capita   = $US 160
- Life expectancy   = 39.7 years
- Literacy (Male)   = 75%
- Literacy (Female)   = 54%
- National debt   = $US 3,000 million
- Annual aid received   = $US 540 million

Social indicators

- Population   = 12 million
- % urban population   = 16%
- Population under 15 years   = 47%
- Population over 65 years   = 3%
- People living with HIV/AIDS = 900,000

Exports

Tobacco (50%)
Sugar (15%)
Tea (15%)
Other agricultural products (15%)

Value $US 500 m per annum

To USA, Germany, UK, South Africa, Japan →

Imports

Food
Oil
Motor vehicle parts
Fertilisers
Other manufactured goods

Value $US 520 m per annum

→ **From South Africa, UK, China, Germany**

(i) Describe the pattern of Malawi's trade.

(ii) Explain why the pattern of trade creates problems for Economically Less Developed Countries such as Malawi.

(iii) *'The economic prospects for Malawi are not promising.'*
Do you agree with this statement?
Explain your answer in detail.

Different types of aid

Many of the Economically More Developed Countries give aid to Economically Less Developed Countries to try to improve their standard of living. This aid takes many forms, as shown in the diagram below:

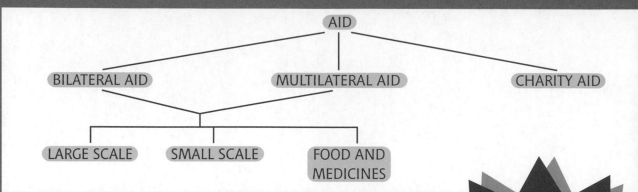

Charity aid

Charity aid is collected in many countries and used to finance small-scale schemes such as improving water supply, educating farmers in new techniques and providing irrigation. At times of emergency, special appeals ask for blankets and money for medicines, which are sent to the victims of famines or natural disasters.

Top Tip
You should be able to name some of the charities which collect money in this way. Examples include SCIAF (Scottish Catholic International Aid Fund), Oxfam and Christian Aid.

Bilateral Aid

Most of the aid sent to Economically Less Developed Countries is sent in the form of **bilateral aid**. This is where one country sends aid directly to another country.

Another major source of aid is sent through international organisations like the United Nations. This is **multilateral aid**. This aid is sent through United Nations organisations such as:

- UNICEF (United Nations Children's Fund, which provides many kinds of aid when young people are affected by disasters)
- WHO (World Health Organisation), which monitors and supports improvements to poor health in all parts of the world), and
- UNESCO (United Nations Education, Science and Cultural Organisation which invests in education programmes to help improve the opportunities for people in mainly ELDCs).

Much of the aid is spent on **large-scale** projects like airports, river dams and power stations.

Increasingly, aid is being given for **small-scale** local projects such as **biogas** plants to produce fuel or tree-planting schemes to reduce soil erosion. Many of these could be described as **self-help** schemes, as money is spent on small projects which local people become involved in. One advantage of self-help schemes is that people learn new skills which will be useful in the years ahead.

Aid is also given in the form of food and medicines, particularly when drought, flooding or other natural disasters strike.

You need to know

You should be able to provide examples of each of the types of aid described. In addition, you should be able to describe the benefits which each brings as well as to explain why there may be problems. To help you, some of these advantages and disadvantages are shown in the table below:

Type of aid	Advantages	Disadvantages
Large-scale aid	• can help many people • may employ many people • encourages long-term industrial development • source country makes money by supplying expertise and materials	• may be tied to other agreements • often not cost-effective • the donor country may get more from the projects than the receiving country
Small-scale aid (self-help schemes)	• many people benefit at local level • low-level technology which is easily understood • repair and maintenance is simple	• little benefit for donor country
Food and medicine	• essential after disaster • improves health	• only useful in the long term if linked with training

Strings attached

It is argued, however, that not all aid is a good thing. Some of the problems include:

- Not all aid may reach the people it is intended for.
- It may destroy the local community; farmers may not continue to work if food is sent.
- The wrong type of aid may be sent to an area.
- Local people may be put out of work, especially if 'experts' are sent in from abroad.
- Aid may only be given if there is an agreement to supply spares or use materials from the donor country.
- Much of the aid given to Economically Less Developed Countries is military: aircraft, weapons, training. This does little to improve the standard of living and may result in unrest or civil war.

Top Tip

Fair Trade is a way of buying products from ELDCs which makes sure that:
- a decent minimum wage is paid to the producer,
- there is no exploitation of children,
- the environment is protected by careful production methods.

Alliances

Different types of alliances

Countries often join together to form **alliances**. They may do this for
- economic reasons
- political or military reasons.

Economic reasons

The European Union is a good example of an economic alliance. The countries which make up the EU have done so to make trading and movement easier among themselves. Goods can pass from one country to another without having to pass through customs or pay taxes.

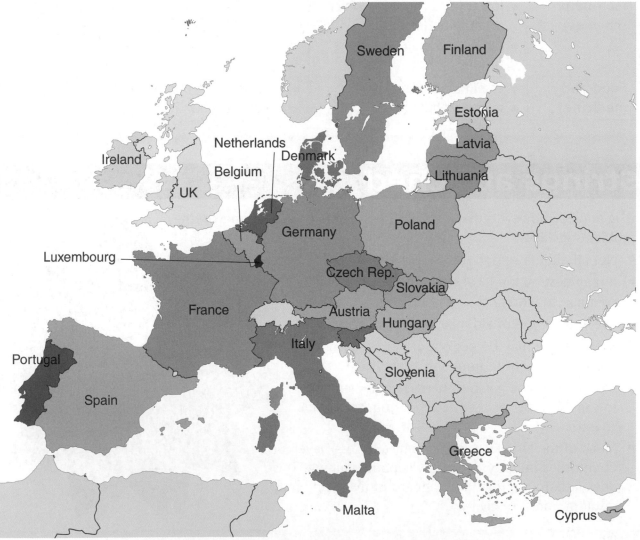

Sometimes countries which produce the same product group together to form another type of economic alliance. The **Organisation of Petroleum Exporting Countries (OPEC)** is an example. Most of the world's oil producers are members and they control the amount of oil exported, thus controlling its price.

Political and military reasons

Countries often form alliances to defend themselves from possible attack. The **North Atlantic Treaty Organisation (NATO)** is one example, where member countries share the responsibility and cost of defence.

An ever-changing world

The existence of alliances, as well as their members, is always changing. Changes in the Eastern European countries, which used to be communist members of the Warsaw Pact, show how quickly things can change.

You will not need to know details about any particular alliance, but should be aware of how:

- certain alliances control large resources, affecting trade and development, and
- some alliances dominate international relations and exert a powerful influence on other countries' actions and borders.

In the exam, you could be asked about the influence of Europe, the USA, or Japan on world events or trade.

Top Tip
Find out what countries are included in the following trading groups:
- NAFTA
- ASEAN
- EFTA
- EU
- OPEC

Answers to exam-style questions

Pages 12–15 Arran OS map

1. **I.**

Arête	990420
Pyramidal Peak	990424
Hanging Valley	977385
Misfit Stream	980405
Corrie	003427

   II. Example:
   An arête is formed when two corries are formed back to back and the back walls of the corries meet in a sharp-edged ridge. While the floor of the corrie is formed by glacial abrasion, the back wall is shaped by the ice ripping or 'plucking' the ice as it moves downhill. Freeze and thaw also keeps the back wall nearly vertical after the ice has melted.

2. There are arguments both for and against building the farm on Glenshant Hill.
   Advantages: High, exposed hill for maximum wind;
   Relatively close to settlements (Brodick) for customers;
   Would help provide 'clean' power for Arran;
   Close to forest tracks for transport of equipment to the site.
   Disadvantages: Would spoil scenery in National Trust area;
   Unnecessary – Arran already has power supply;
   Eyesore – seen from both mountains and the coast;
   Expensive to transport all necessary equipment to this location.

3. Either agreeing or disagreeing is possible.
   If agreeing, the following points could be expressed:
   i. Already a tourist area – many facilities (e.g.hotel at 010359) already in place – all year round employment;
   ii. Superb scenery – would attract many tourists bringing income to the area;
   iii. Ferry to Brodick (0236) – easy to reach and would employ locals;
   iv. Unemployment in Scottish islands is high – this would provide much-needed income for locals and the area;
   v. Any new developments could be well hidden in valleys (e.g. Glen Rosa) or in forest (0038).
   If disagreeing, the following points are valid:
   i. This is an area of outstanding beauty – large-scale development would destroy the island;
   ii. Ski-lifts, public paths, car parks, new hotels, visitor centres, resort features would have a negative impact on the environment;
   iii. Much of the investment would be by non-local companies – any income would not stay on the island; incomers would be employed and the island would not benefit;
   iv. Many people have moved to Arran for peace and quiet – this development would destroy their quality of life;
   v. Crime, pollution, vandalism, and noise might increase;
   vi. Many traditional features (e.g. Brodick Castle and ancient monuments

– 002363) might lose their appeal.

4. There could be conflict between any of the groups mentioned. For example:
   Traditional sportsmen would be unhappy at hill walkers or climbers disturbing their targets, particularly at certain times of the year; walkers could also affect the breeding of certain animals or birds.
   Farmers are not always very comfortable with tourists who don't always understand the country code and might disturb animals, leave gates open, walk through crop fields or leave litter behind.
   The forestry groups need access to their trees for trimming and felling and this could restrict access for hikers or other tourists.

5. The Glen Rosa Water has its source high in the Arran mountains at a height of around 600 metres. Flowing from the floor of a corrie (Fionn Choire), it rapidly plunges to the foot of Glen Rosa, a U-shaped, glaciated valley running in a south-easterly direction. In 9838, the valley becomes even narrower, turning east, and the Glen Rosa Water begins to meander in its flatter floor. In 9937 and 0037 there is evidence of the river's floodplain, and the river turns south-east once again before widening and entering the sea in Brodick Bay. There are many tributaries of the Glen Rosa Water, those joining in the U-shaped valley running from the steep slopes, sometimes hanging valleys (9738) and the most significant tributary is the Shurig River, with its confluence at 002369.

6.
Glen Rosa	D
Creag Rosa	E
Cnoc Breac	C
Beinn Nuis	A
Coire nam Meann	B

7. **W** Hill sheep farming, possibly hunting (in season), hill walking. The land is too high for cultivation or forestry. These are only likely uses. The land is relatively flat, likely to be marshy.
   X Commercial forestry. Although land is steeply sloping, it is easy to access and still within growing range for coniferous trees.
   Y Pasture is the most likely use, as ground is lower and sheltered. Sheep or cattle would be suited to the lower altitude.
   Z This is an area of parkland and open forest around Brodick Castle.

8.
A	Coire Daingean
B	Beinn Chliabhain
C	Goatfell
D	The Saddle

Pages 15–17 Hamilton OS map

1. 7255 – evidence includes:
   - converging roads;
   - bus/rail stations;
   - inner ring road;
   - public building – town hall.

2. Earnock has characteristics of a late 20th-century housing area, with roads in crescents and cul-de-sacs, often following the contours and no through main roads. It is also built at the edge of Hamilton, typical of modern developments.

Orbiston, on the other hand, is more typical of mid 20th-century housing, with a more geometric street pattern. It is also closer to the centre of Bellshill. A main road runs through the estate.

3. There are a number of possible gathering techniques including observation. This would require visits to the area and the production of annotated sketches or photographs which could be used to highlight different ages of buildings, remnants of industrial/residential areas.
   Old maps of the areas could also be examined to investigate changes in the landscape – a series of maps could show the changes over a number of years.
   Another method of gathering information would be to interview local residents who had memories of the past. This provides graphic and social evidence which can complement evidence gathered in different ways.

4. A number of advantages could be identified, including:
   - The industrial park is close to a workforce in Blantyre, Hamilton and East Kilbride; this will provide jobs in an area with a traditionally high unemployment level.
   - Lanarkshire was a traditional industrial area with many people working in heavy industry. This can be seen in the old coal spoil tips (685545 and 675588). Technology parks are designed to attract 'sunrise' or light industries which will boost the local economy;
   - East Kilbride was a New Town built in the mid 20th century. Areas around the New Towns often felt disadvantaged by their comparative lack of attraction, so local businesses would welcome this technology park; particularly those in Blantyre and Hamilton.
   - The development of new jobs will bring more money into the local community: many small businesses such as shops and services will prosper as more people have money to spend.

5. There are a number of similarities between the land use model and Hamilton. There is evidence of CBD functions in 7155 and 7255 (Town House, converging roads, bus/train stations, ring road, County Buildings, churches) and, as in the model, older housing mixed with industry is seen close to this (unplanned streets in 723559). Further from the town centre there is evidence of more modern housing – 717545 is mid-20th century style with geometric patterns and 706537 shows more modern housing with crescents, cul-de-sacs and contour-following streets. There is also evidence that the road pattern is important as the B7071 is beside parkland, a race course and industrial buildings (7156). There is also evidence that the A723 retains older building styles from the CBD at 723553 to 716536.
   There are also differences between the model and Hamilton: the CBD is far from the geographical centre of the town, only 0.5 kilometres from the town boundary to the north west. The area occupied by the CBD seems to be well spread out, possibly into two areas around 723554 and 716557. Unlike the model, Hamilton seems to merge

into adjacent towns such as Blantyre and Motherwell.

6. The River Clyde in the map extract is flowing in a north-westerly direction. It is in its lower stage, flowing in a wide floodplain and meandering extensively. In 7356 the floodplain is nearly 1 kilometre wide and the river is meandering. Following its confluence with the River Avon (737560), the River Clyde is artificially straightened to run alongside Strathclyde Loch. At this point, the river valley floor is nearly 2 kilometres wide. Between Bothwell Bridge (712576) and Bothwell Castle (687594) the river runs in a steep-sided valley, generally following a northerly route before once again entering a wide floodplain. In 6761 and 6861 there is a very large meander and a river island which indicates deposition.

7. This site offers both advantages and disadvantages as a location for a country park. On the plus side, it is an area on the floodplain of the River Clyde which would be prone to flooding and improved drainage of the land would be welcomed. There is flat land which is suitable for a range of activities such as jogging and cycling, and the man-made loch is suitable for water sports. The site is also suitable for other activities such as forest trails, golf and horse racing, as shown on the map extract. There are, however, a number of disadvantages of the site. The flat land which makes it so suitable for so many activities also means it is very prone to flooding and the installation of drains and run off channels would be expensive. The closeness of large centres of population (Hamilton, Motherwell, Bellshill) means that pollution and overuse might cause problems.

8. There are a number of techniques which could be used to find out if Bothwell is a dormitory settlement.
Pupils could carry out traffic surveys at rush hour on local roads to find out the number and direction of traffic movement. The destination and frequency of bus journeys would also give valuable information.
Another technique would be to use a questionnaire to ask local people where they worked. This could be done throughout the day (including rush hour) at local bus stops and Hamilton West and Uddingston railway stations. This technique could be developed by asking a cross-section of the people in Bothwell the same questions. A cross-section might be found during a weekend; carrying out a survey during a weekday would give a biased viewpoint.

9. Muirmains Farm is located at a height of approximately 180 metres on a sloping site to the south-west of Hamilton. The land has a north-easterly aspect. The most likely type of farming for the farm would be dairy farming. The combination of altitude, relatively high rainfall in the west of Scotland and hilly land mean grass is the only viable crop. The closeness to a large market in Hamilton, East Kilbride and nearby Glasgow suggests milk production is the most likely product for the farm.

10. The River Clyde in Area X is both enclosed within a deep valley and flowing through a wider valley with a narrow floodplain. In the southern part of Area X, the steep valley has limited potential for varied land use. The valley sides are wooded, with tracks suggesting recreational use. There is evidence (weir at 696583) of former industrial use, using the river for water power. Some settlement reaches the river's edge on the less steep, western side (6958). The steep slopes of the valley were well suited to the building of a castle at 687593. In the northern half of Area X, the town of Uddingston has reached the river's edge in 6960. The existence of a floodplain has restricted riverside developments to the west of the river.

11. The most likely answer would be to question the proposal on the grounds that the new development would be sited on the floodplain (lack of contours) of the River Clyde, and as such would be prone to periodic flooding when the river overflows. Householders would find it difficult to get their properties insured.
The development would also be very close to a sewage works and a crematorium (6761) which might be seen as unpleasant. The adjacent motorway intersection would create noise and air pollution, another negative factor.

12. The building of the M74 created a number of problems for engineers. It had to be built over a number of existing roads and railway lines (7060) and was built cutting through a residential area (6961). This would require expensive bridge-building and soundproofing. At 710595, the road was cut through higher land, requiring the building of bridges over the motorway. From 711590 southwards, the M74 was built on an embankment to raise it above the floodplain of the River Clyde.

13. I. It is easier to argue in favour of the retail park, as the following points could be expressed as advantages:
- Available flat land at the edge of the town;
- Close to existing town centre for customers;
- Beside major roads (M74, A723) for easy access for both customers and supplies;
- Land available does not appear to be used for anything else.

II. Research students could use techniques such as
- Questionnaires to local shoppers to find out their feelings on the new development;
- Questionnaires of local shopkeepers to find if business had improved/deteriorated as a result of the development;
- Shopkeepers in the new shopping area and those in the traditional shopping centre could be interviewed;
- Traffic/parking surveys could be carried out to find out how much use was made of the retail park.

Page 29

A gorge is formed by river erosion. When a river flows over high ground it will find it difficult to carve out a valley if the underlying rock is resistant. If, however, the rock below the resistant rock is softer, when a waterfall is formed (often caused by a crack or fault in the harder rock), the falling water will erode the softer rock, gradually undermining the overlying rock until it collapses. This material will be swirled around by the river falling over the waterfall, causing the softer rock to erode more and more by a process of attrition and corrasion. Over a period of time, this process is repeated many times and the waterfall gradually retreats upstream, leaving a deeply cut scar or gorge in the landscape.

Page 37

An example of glacial erosion is a corrie. During the ice age, snow gathered in hollows on the upper slopes of mountains and, under pressure, turned to ice. This ice moved downhill. Abrasion occurred at the sole of the glacier, gouging out a deeper hollow. At the same time, plucking at the back wall of the hollow steepened it and produced a corrie. After the ice melted, the typical corrie shape could be seen: a deep hollow in a hillside with a lip, often containing a lake or corrie lochan.
An example of glacial deposition is an esker. An esker is formed by rivers flowing inside or at the base of a glacier. As with all rivers, bedload (rocks, pebbles and grit) are deposited below the river, which creates a larger pipe in which to flow. When the ice melts, a long, often twisting bank of debris is left on the earth's surface, following the route of the original river.

Page 41

1. Explain in detail what will happen to the weather in Newcastle over the next twenty-four hours.
At present the weather in Newcastle is very cloudy with fairly strong south-west winds. The cloud level is low and persistent rain will last for several hours. Air pressure, already low (988 mb) will continue to fall. The temperature will rise over the next few hours and the winds will gradually veer to the west. This will continue for a short time but a period of heavy rain will accompany a drop in temperature. Following this there will be a shift of the wind to the north-west then north, and the weather will become brighter, with sunny periods punctuated by heavy, squally showers. At this time, air pressure will rise continually, reaching around 996 mb.

2. Plot A is the most likely answer. Edinburgh is in an area of high pressure with light winds from the west. There will be little cloud and temperatures should be high in June. It is unlikely that there will be any rainfall.

Page 45

The first climate graph shows a typical Mediterranean climate. Average temperature range from 12°C in January (winter) to 21°C in July and August (summer). This gives an annual temperature range of 9°C. The average temperature never falls below freezing. Rainfall figures are low in summer but fairly high in winter (60–70mm in December and January). The summer period (May to September) has very little rainfall – typical of a Mediterranean climate.

The second climate graph is typical of a hot desert climate. Temperatures are always high, ranging from 17°C in January (winter) to 35°C in August (summer). This gives an annual range of 18°C. Rainfall is very low throughout the year, ranging from total drought in summer to a very low 20mm in January. The total annual rainfall is only 70mm, typical of a desert climate.

Page 49

There are several potential conflicts between users in a national park such as the Lake District. Farmers would normally prefer their fields to be protected from walkers and ramblers so activities such as grazing, lambing or harvesting are not disturbed. Tourists may also leave litter lying which is harmful to livestock, or leave gates open which can be dangerous. Hotel owners and shopkeepers will generally support tourists as they will provide income. Many house builders will benefit if tourism is important, as the area may become attractive as a retirement option or for people to buy holiday homes. This will improve the market for new or renovated houses. National bodies like the Forestry Commission and Water Board own large areas of countryside and are generally keen to encourage visitors to learn about the area and their activities. They may also have concerns, however, about people polluting water supply or damaging small trees. Conservationists, while keen to provide opportunities for people to understand the countryside, may have concerns about damage to the environment, additional building on rural land, or the fuel used travelling to the national park.

Page 61

Building large retail parks will offer a number of advantages for both retailer and shopper. Land values often tend to be lower at the edge of the city, and there is often more flat land available for large complexes with car parking space. Ring roads/by-passes mean supplies can be brought in easily. There are disadvantages, however. Traditional city centre stores may lose custom and be forced to downsize or close, meaning a loss of services and the 'buzz' of the town centre.

Shoppers may also have mixed feelings: those with cars or access to bus services may enjoy the large stores and range of shops available, but others may miss the traditional shopping experience of the city centre.

There are also issues about the loss of green belt or farmland around urban areas, and conservationists often protest that too many people use too much petrol in travelling unnecessary distances to shop in these retail parks.

Page 70

i. Between 1965 and 2005 there have been many changes on Holm Farm. Increasing use of machinery such as tractors and combine harvesters has meant that removing fences to make larger fields is more efficient, so the number has more than halved. The Common Agricultural Policy of the European Union has allowed farmers to maximise profits by concentrating on a fewer number of crops, so there is less variety on the farm (2 crops plus grass for fodder instead of 3). In order to maintain a decent lifestyle, many farmers have felt the need to diversify. This means replacing traditional crops with new ways of making an income. On Holm Farm, the farmer has created holiday cottages from the old workers' houses (no longer needed because of increased mechanisation which reduced the number of farm workers needed). This will generate income as will turning two fields into a golf range. Building a farm shop which sells produce from the local area will also bring in money from both local people and tourists. Planting trees for profit is a long-term investment which puts poor quality land to good use.

ii. There are a number of techniques which could be used to record the relationship between land use and relief and drainage. One way would be to use field sketches and/or photographs to record the landscape; these could then be annotated to show the

relationship. Another method would be to use instruments such as a clinometer and ranging poles to record the different slopes. This information could produce a transect on which details of land use and drainage could be marked.

iii. Many farming areas in Britain are suffering from rural depopulation because there is often a lack of opportunity for (particularly) young people in rural areas and many leave to further their careers in towns and cities. This leaves an ageing population which becomes a declining population (fewer children born). This in turn causes problems for small businesses and services which may close, making the area even less attractive to those living there. This encourages even more people to leave.

Page 83

1. i. In the selected countries, there is a clear difference between the countries of the North and the South.
    Countries in the North have a high proportion of their population living in urban areas, ranging from 92% in Australia to 66% in Japan. In the South, the level of urbanisation is generally much lower, as low as 17% in Malawi. Brazil seems to be an exception – its figure of 84% would be more typical of a country in the North. Africa and South Asia have the lowest level of urbanisation.

    ii. The North contains many countries which are Economically More Developed Countries. These are wealthier countries which make much of their income through the production of manufactured goods and services. These industries need specialised working conditions which are typically found in built up areas. Only a relatively small proportion of their wealth comes from agriculture, so there is a great incentive for people to live in towns and cities.
    The South, on the other hand, contains most of the world's Economically Less Developed Countries. These countries are much poorer, often depending on the export of raw materials (e.g. timber, coffee, metal ores) for their income. Manufacturing industries and services are relatively underdeveloped. The majority of the population live in rural areas, depending on the land for their existence. The level of urbanisation is rising, but still lags behind the North except in a limited number of countries such as Brazil.

2. Although rural poverty is still very common, there is a mass migration to urban areas in search of employment and a better quality of life. There are both 'push' factors which encourage people to leave their home area and 'pull' factors which attract them to the urban areas.
    The rural areas are often overpopulated with insufficient food to feed families. There are few well-paid jobs and poverty is common. Poor health care and education mean disease is high and prospects poor for those remaining in these areas.
    In hope of a better life, many move to urban areas where they hope there will be paid work and better prospects for their families. Unfortunately, this is seldom the case, as cities are overcrowded and health care and education is considerably overstretched. Many live in shanty towns with poor sanitation and cannot find work. People continue to move, however, mainly

because there is hope, and a perception that life in the city offers opportunities which are certainly not found in the rural areas.

3. Afghanistan has a population pyramid which is typical of an Economically Less Developed Country. There are many young people as birth rate is very high, common in countries where infant mortality is high (disease, malnutrition). There are very few people living to old age because of poor health care and tough lives for the majority of workers.
    The Netherlands, on the other hand, has a much lower birth rate (high living standards and extensive family planning) as well as excellent health care.

Page 86

i. Malawi's trade is typical of a Economically Less Developed Country. Its exports are mainly raw materials or unprocessed goods which are largely sold to Economically More Developed Countries such as Germany or Japan. The imports are processed goods which Malawi has neither the resources nor expertise to produce.
    The value of imports is greater than exports, leading to a trade deficit.

ii. This pattern of trade creates difficulties because selling unprocessed goods means there is no 'added value' and only the minimum income can be made. The value of these products is also liable to rapid change meaning a fall in price can severely reduce a country's income. The power in setting prices for goods rests mainly with the rich, or Economically More Developed Countries; this will not normally work to the advantage of countries like Malawi. Malawi also suffers a trade deficit. This means it is in debt, reducing the amount the country can spend on investments in manufacturing industries, and services such as health and education.

iii. Most people would agree with the statement because the poor balance of trade and the type of export which Malawi has are unlikely to lead to an increase in wealth. Coupled with a rapidly increasing population, typical of an Economically Less Developed Country, there seems to be little chance of a healthy economic future. It is likely that Malawi will have to negotiate loans from wealthy countries or the World Bank, both to balance the books and to find money for investment in industries or services which will help it out of its difficulties. Loans and aid come with ties – this could affect the decisions Malawi's government can make.

Catchword sheet

See page 6 for instructions on completing this sheet.

Catchword sheet	Topic:
Catchword	**Meaning**

Index